PEAS

CULTIVATION, VARIETIES AND NUTRITIONAL USES

FOOD SCIENCE AND TECHNOLOGY

Additional books in this series can be found on Nova's website
under the Series tab.

Additional e-books in this series can be found on Nova's website
under the e-book tab.

NUTRITION AND DIET RESEARCH PROGRESS

Additional books in this series can be found on Nova's website
under the Series tab.

Additional e-books in this series can be found on Nova's website
under the e-book tab.

FOOD SCIENCE AND TECHNOLOGY

PEAS

CULTIVATION, VARIETIES AND NUTRITIONAL USES

AIDEN M. COMSTOCK

AND

BRAYLEN E. LOTHROP

EDITORS

Nova Science Publishers, Inc.

New York

For permission to use material from this book please contact us:
Telephone 631-231-7269; Fax 631-231-8175
Web Site: http://www.novapublishers.com

NOTICE TO THE READER

Additional color graphics may be available in the e-book version of this book.

Library of Congress Cataloging-in-Publication Data

Peas : cultivation, varieties and nutritional uses / editors, Aiden M. Comstock and Braylen E. Lothrop.
      p. cm.
 Includes bibliographical references and index.
 ISBN 978-1-61942-866-9 (soft cover)
 1. Peas. 2. Peas--Varieties. 3. Peas--Utilization. 4. Nutrition. I. Comstock, Aiden M. II. Lothrop, Braylen E.
 SB343.P43 2011
 635'.656--dc23
                     2012000574

Published by Nova Science Publishers, Inc. † New York

CONTENTS

PREFACE

In this book, the authors present topical research in the study of the cultivation, varieties and nutritional uses of peas. Topics discussed in this compilation include the enhancement of the nutritional value and functional properties of yellow pea protein via membrane processing; the volatile flavour profile of 24 pea cultivars using head space solid phase microextraction gas chromatography; regulation of nodule developments in Pisum sativum L.; developing fall-sown pea cultivars as an answer to the challenges of climatic changes; and the chemical, molecular structure, biodegradation and nutritional characterization of peas.

Chapter 1 – Pea proteins (*Pisum sativum*) with, reasonably, well balanced profile of amino acids and storage globulins play important roles in animal and human nutrition and have been widely used to enhance the texture of many foodstuffs. Despite the nutritional potential of peas as an economic source of proteins, carbohydrates, vitamins and some minerals the utilization of this legume has been limited, due to the presence of certain antinutritional factors such as phytic acid, polyphenols including condensed tannins, protease inhibitors (trypsin and chymotrypsin), amylase inhibitors, lectins, oligosaccharides and saponins. Among these antinutritional factors, phytic acid has been found as a strong chelator of mineral nutrients, such as Ca, Zn and Fe. The complex of phytic acid and mineral elements, in the form of phytate, could cause a noticeable reduction in bioavailability of these nutrient elements, and a consequent public health problem of iron and zinc deficiency, particularly, for those populations living on consumption of legumes as the main source of nutrients. In addition, animals fed with feed of high phytic acid content release enough phytate to be a threat for contamination of water. Phytic acid has also an inverse effect on solubility of protein through the formation of insoluble complexes at pH values below protein isoelectric point. This effect will then limit functional characteristics and exploitation of the legumes proteins in foodstuffs. Attempts to reduce antinutritional factors and increase the utilization of legumes have employed a wide range of processing techniques such as germination, dehulling, cooking, roasting, autoclaving, fermentation, extrusion cooking and recently membrane processing. Membrane processing, membrane-based pressure driven process, has been widely used to purify, separate, and concentrate colloids and high molecular weight materials in solution. In addition, removing particulates, bacteria, and pyrogens, as well as recovering valuable ingredients in the food, chemical, and pharmaceutical have been found as applications of membrane processing. Furthermore, isolate form of plant proteins have widely been employed to improve the nutritional quality and functional properties of the food products. Consequently, the idea of using membrane processing to purify pea protein and exploit its uses cover the main proposes of this chapter. Moreover, the results of our recent research study on isolation of pea protein via membrane processing including characterization and uses will be included to bring novelty to this chapter.

Chapter 2 – In this study, the volatile flavour profile of 24 pea cultivars were identified using head space solid phase microextraction gas chromatography coupled with mass spectrometry. Cultivars from several breeding programs were utilized with the purpose of understanding the genetic diversity of the cultivars in relation to differences in their flavour compounds. Furthermore, the effect of market class, cultivar, location, and crop year on the flavour characteristic of peas was evaluated. Cultivars were selected from 4 market classes of peas (i.e., yellow, green, dun, and marrowfat) grown in four locations (i.e., Sutherland (SUT); Meath Park (MPK); Pasqua (PAS); Wilkie (WIL) in Saskatoon, Canada in different crop years. Significant differences ($P < 0.01$) in volatile flavours were observed between field pea cultivars and the concentration of flavour compounds was significantly affected by the parameters studied. The 2008 crops had higher TVC, alcohols, ketones, pyrazines, and hydrocarbons than those from the year of 2009. In contrast, the 2009 crops had higher concentrations of aldehydes, sulfur compounds, esters, and terpenes. Overall, Cooper and Rambo had the highest mean value of TVC, whereas Kaspa had the lowest mean value of TVC. Peas grown in WIL and MPK had higher TVC compared to those from other locations. The highest mean value of TVC was observed in peas from the marrowfat-market class, whereas peas from the dun-market class had the lowest TVC. The lowest value of alcohols, aldehydes, ketones, sulphur compounds, hydrocarbons and pyrazines was found in the marrowfat-market class. In contrast, the dun-market class had the highest value of these compounds except for ketones and alcohols. Ketones were found in the highest amount in yellow- and green-market classes, and alcohols were observed at the highest level in green-market class. 2-Ethyl-1-hexanol, hexanal, 2-butanone, dimethyl sulfide, styrene, 2,3-diethyl-5-methyl pyrazine, ethyl acetate and hexanoic acid,methyl ester were the most abundant flavour compounds found in peas grown in both crop years. As the taste and flavour of peas could be affected by varying concentrations of different volatile compounds, these findings could be useful in the identification of specific cultivars for different food applications.

Chapter 3 – The symbiotic interaction of legume plants with soil bacteria of the genus *Rhizobium* results in the formation of new organs on legume roots – nodules, where symbiotic nitrogen fixation takes place. Symbiotic nodules are formed as a result of dedifferentiation and reactivation of cell division of cortical root cells. A few data are available regarding mechanisms regulating nodule meristem development and function. In our work the authors focused on meristem development of indeterminate-type nodules in *Pisum sativum* L. According to recent data there are common features in regulation of nodule meristem and regular apical meristems. In this chapter the authors will briefly review known mechanisms underlying plant apical meristems development and maintenance with special attention to homeodomain transcription factor of WOX family and CLAVATA-related receptor complexes. Based on analysis of pea nodule development the authors will discuss the role of WOX and CLAVATA-related genes in nodule organogenesis.

Chapter 4 – Pea is considered rather well adapted to wide temperature ranges, with seedlings able to survive even -20 °C. From a physiological viewpoint, pea becomes tolerant to frost if first exposed to low non-freezing temperatures, causing the so-called cold acclimation. Delayed floral initiation helps some forage pea genotypes to escape the main winter freezing periods, as susceptibility to frost increases during the transition to the reproductive state. The oldest winter pea cultivars carry the dominant allele, *Hr*, although some bear *hr*. They are generally characterized by prominent winter hardiness and a long growing season, from sowing in early October until either cutting for forage production in late May or harvesting seeds in mid-July. The average forage yields in the winter forage pea cultivars often exceed 45 t ha^{-1} of green forage, 9 t ha^{-1} of forage dry matter and 1700 kg ha^{-1} of forage crude protein. Modern dry pea cultivars have advanced winter hardiness and enhanced dry grain production. They are already in use in other temperate regions in both Europe, especially France, and the USA. One of the strategic advantages of fall-sown dry pea cultivars of recent release is their significantly improved earliness. These cultivars are regularly at least one week earlier than winter barley, providing many farmers with the novel opportunity of not having to choose between pea and cereals, since many have only one combine harvester available and give priority to their cereals. Furthermore, fall-sown dry pea cultivars may have increased grain dry matter crude protein content and it is possible to merge winter hardiness and low content of anti-nutritional factors. Low thousand seed weight, not exceeding 200 g, and a population density of 75-80 plants m^{-2} provide inexpensive sowing. All these outcomes should result in an increased area and production of dry pea in many temperate regions. In the end, growing winter-hardy pea cultivars also mean a significant shift into the wetter half of the year and thus mitigating more and more prominent and unpredictable effects of spring droughts, demonstrating an efficient answer to the challenges of climatic changes.

Chapter 5 – The Latin name for pea seeds is *Pisum sativum*. The peas are also called marrowfat pea, small blue pea, dry pea, arveja, muttar etc. Peas have attracted attention as components of food and feed for human and animals as protein and energy supplements mainly because they usually have a particularly high protein and starch contents. Peas are also as one of health food for human due to fact that they are high in fiber, vitamins, minerals and lutein etc. Although pea seed peptide fractions have less ability to scavenge free radicals than glutathione, greater ability to chelate metals and inhibit linoleic acid oxidation.

In: Peas   ISBN: 978-1-61942-866-9
Editors: A. Comstock and B. Lothrop   © 2012 Nova Science Publishers, Inc.

Chapter 1

ENHANCING NUTRITIONAL VALUES AND FUNCTIONAL PROPERTIES OF YELLOW PEA PROTEIN VIA MEMBRANE PROCESSING

Ali R. Taherian, Martin Mondor*
and François Lamarche
Food Research and Development Center of Agriculture
and Agri-Food Canada, St-Hyacinthe, Quebec

ABSTRACT

Pea proteins (*Pisum sativum*) with, reasonably, well balanced profile of amino acids and storage globulins play important roles in animal and human nutrition and have been widely used to enhance the texture of many foodstuffs.

Despite the nutritional potential of peas as an economic source of proteins, carbohydrates, vitamins and some minerals the utilization of this legume has been limited, due to the presence of certain antinutritional factors such as phytic acid, polyphenols including condensed tannins, protease inhibitors (trypsin and chymotrypsin), amylase inhibitors, lectins, oligosaccharides and saponins.

Among these antinutritional factors, phytic acid has been found as a strong chelator of mineral nutrients, such as Ca, Zn and Fe. The complex of phytic acid and mineral elements, in the form of phytate, could cause a

---

* E-mail address: alitaherian@agr.gc.ca; Tel: 450-768-3329 (Corresponding Author)

noticeable reduction in bioavailability of these nutrient elements, and a consequent public health problem of iron and zinc deficiency, particularly, for those populations living on consumption of legumes as the main source of nutrients. In addition, animals fed with feed of high phytic acid content release enough phytate to be a threat for contamination of water. Phytic acid has also an inverse effect on solubility of protein through the formation of insoluble complexes at pH values below protein isoelectric point. This effect will then limit functional characteristics and exploitation of the legumes proteins in foodstuffs.

Attempts to reduce antinutritional factors and increase the utilization of legumes have employed a wide range of processing techniques such as germination, dehulling, cooking, roasting, autoclaving, fermentation, extrusion cooking and recently membrane processing. Membrane processing, membrane-based pressure driven process, has been widely used to purify, separate, and concentrate colloids and high molecular weight materials in solution. In addition, removing particulates, bacteria, and pyrogens, as well as recovering valuable ingredients in the food, chemical, and pharmaceutical have been found as applications of membrane processing.

Furthermore, isolate form of plant proteins have widely been employed to improve the nutritional quality and functional properties of the food products. Consequently, the idea of using membrane processing to purify pea protein and exploit its uses cover the main proposes of this chapter. Moreover, the results of our recent research study on isolation of pea protein via membrane processing including characterization and uses will be included to bring novelty to this chapter.

1. NUTRITIONAL FACTORS OF PEA CULTIVARS

Legume seeds, in general, play an important role in human and animal nutrition. Dry legumes are main ingredients of diet in many parts of the world and have been considered as the most significant food sources for people of low incomes due to their high protein content (Makri, Papalamprou and Doxastakis, 2005). Although legumes have historically been utilized mainly as whole seeds the utilization of legumes in other forms (e.g. like flour, concentrate, isolate) has grown in recent years (Doxastakis, 2000). In addition, the use of plant protein products in food as functional ingredients to improve the stability and texture as well as the nutritional quality of the product or for economic reasons is very extended. Nevertheless, these applications in the food trade are almost limited to proteins from soybean seeds, whereas other vegetable proteins are less used. Among these are those from lupins (Lupinus albus L.), peas (Pisum sativum L.) and broad beans (Vicia faba L.), that are extensively grown in different parts of the world and, in particular, in the Mediterranean region. Their high protein content (34.7% for lupin, 23.4% for pea and 32.5% (w/w) on d.m. for broad beans and their well-balanced amino-acid composition makes them important sources of protein, with potential to add in various products as novel food ingredients (Dervas, Doxastakis, Hadjisavva-Zinoviadi, and Triantafillakos, 1999; Doxastakis, 2000). However, dry peas (*Pisum sativum*) have the largest production volume of all special crops in Canada. In addition, the yellow pea is the most widely seeded and produced, with approximately 40 varieties registered in Canada, while the newest type, the green marrowfat, has two registered varieties (Agriculture and Agri-Food Canada, Pulses and Special Crops, 2007).

The proteins present in legume seeds can be broadly classified into metabolic proteins, which are involved in normal cellular activities, and storage proteins which are synthesized during seed development. Of the storage proteins, globulin constitutes a major proportion of the legume seed proteins and its limitations in human and monogastric nutrition are well known. The storage proteins, 7S (conglycinine) and 11S (glycinin), are the principal components of soybean proteins, whereas 7S (vicilin) and 11S (legumin) are the principal components of pea proteins (Derbydhire et al., 1976). A gel of 7S has been shown to be harder than that of the 11S from soybean proteins. On the other hand, pea proteins, have been reported to form heat-induced gels with purified vicilin, but not legumin (Bora, Clark, and Powers, 1994). However, in general, the protein quality of legumes suffers from low levels of essential amino acids, namely methionine, cysteine and tryptophan (FAO, 1970).

The characterization of peas as legume crops is based on relatively high contents of crude protein and metabolisable energy. Igbasan and Guenter (1996) reported that the crude protein and metabolisable energy contents of pea cultivars range from 160 to 320 g per kg. Lallés (1993) indicated that pea contains less fat but the starch content is high which could be regarded as a limiting factor for exploiting pea flour. In terms of amino acids, pea appears to possess fairly high contents of essential amino acids. Pea protein is richer in lysine than soybean protein. Both display low and high contents of methionine and arginine respectively. However, their threonine content is lower than that of cow milk protein. Study by Gatel and Grosjean (1990) also classified the European pea cultivars (*P. sativum hortense* and P. *sativum arvense*) according to their grain characteristics into white peas (seed coat white/yellow, round and smooth), marrowfats (seed coat blue/green, large, flat round, and wrinkled), small blues (seed coat blue/green, small, round and smooth) and large blues (seed coat blue/green, large, round and smooth). They indicated that all these groups have spring and winter cultivars. White pea seeds are primarily of use in animal feeds whereas the main uses of other types are for human consumption. *P. sativum arvense ("coloured"* or "fodder" peas) have dark-coloured flowers and a dark brown seed coat; they are mainly produced in the nordic countries of Europe and are primarily fed as the whole plant to ruminants. They also reported that essential amino acid contents of European pea are about 18.5 g lysine, 6.2 g methionine plus cystine, 9.6 g threonine and 2.0 g tryptophan kg^{-1} dry matters (DM). However, as compared with soybean meal protein, pea protein contains more lysine, a similar quantity of threonine, but less sulphur amino acids and tryptophan.

In another study by Gatel (1990), the average crude protein content for smooth peas was reported to be about 255 g kg- 1 dry matter (DM). It was also added that, for total protein of peas, main protein fractions were 60% globulins (legumin and vicilin) and 20% albumin. The nutritional factors of different grain legumes such as feed pea, lupin, cow pea and rice bean have also been reported in studies by Cruz-Suarez et al., (2001); Davis et al., (2002); Bautista-Teruel et al., (2003); Sudaryono et al., (1999); and Eusebio, (1991) respectively. However, the protein content of peas, largely, depends upon the type of process applied to extract the protein (Sunday et al., 2001; Kumaraguru Vasagam, Balasubramanian, Venkatesan, 2007; Uwaegbute et al., 2000; Rivas-Vega et al., 2006; Mubarak, 2005).

2. ANTI-NUTRITIONAL FACTORS OF PEA CULTIVARS

Peas contain a number of bioactive substances including enzyme inhibitors, lectins, phenolic compounds, phytates, oligosaccharides, and saponins that play metabolic roles in humans or animals (Campos-Vega, Loarca-Pina and Oomah, 2010). The effect of these substances may be regarded as positive, negative, or both (Champ, 2002; Campos-Vega, Loarca-Pina and Oomah, 2010). However, since these natural occurring substances may interfere with nutrient availability, they have been in most studies considered as antinutritional factors. For example, enzyme inhibitors can reduce protein or carbohydrate digestibility and lectins may result in reduction of nutrient absorption. Trypsin inhibitors strongly inhibit trypsin activity reducing the digestion and absorption of dietary protein. Lectins, also known as phytohemagglutenins, are capable of agglutinating red blood cells. Some lectins have been associated with growth depression in experimental animals (Chung, Wong, Wei, Huang, and Lin, 1998; Sandberg, 2002). Protein digestibility and mineral bioavailability of peas can also be affected, adversely, by phenolic compounds. For example, tannins inhibit the digestive enzymes and thereby lower digestibility of most nutrients, especially protein and carbohydrates.

Owusu-Ansah and Mc Curdy (1991) stated that peas contain lipoxygenase in small quantities which is important in processing and/or utilization of pea fractions. Lipoxygenase contributes both desirable and undesirable effects in foods. When lipoxygenase is present in legume flour at levels of 0.5% to 1.5%, it enhances the performance of flour in baked products. However, the presence of lipoxygenase in raw legumes is associated with the development of off-flavors during storage. Phytic acid can diminish mineral bioavailability and decrease functionality of pea protein (Sandberg, 2002). Liu et al (2005) reported that phytic acid presents in legumes is a strong chelator of mineral nutrients, such as Ca, Zn and Fe. The complex of phytic acid and mineral elements, in the form of phytate, leads to a marked reduction in bioavailability of these nutrient elements, and a consequent public health problem of iron and zinc deficiency. Phytic acid was considered as an antinutrient mainly due to its ability to bind essential dietary minerals as well as proteins and starch, and to consequently reduce their bioavailability in humans.

Furthermore, the presence of the α-galactosides raffinose, stachyose and verbascose in legume seeds are the other reasons for low consumption of these products (Reddy et al., 1985; Norton, 1991; Wang, Phillippy, 2003; Domoney, Hedley, Casey, and Grusak, 2003; Khattab and Arntfield, 2009).

3. INTRODUCTION TO MEMBRANE PROCESSING: EFFECT ON PURIFICATION AND ISOLATION OF PEA PROTEIN

Ultrafiltration is a membrane filtration process in which a transmembrane pressure forces a solution against a semipermeable membrane. Water and solutes feed stream below the nominal weight cut-off of the membrane are allowed to permeate through the membrane pores into a permeate stream, while the larger feed components are retained by the membrane (Figure 1). These retained components accumulate in the retentate stream and are concentrated. By implication, the ultrafiltration process can be used for purification of solution (removal of some unwanted components from a liquid stream), separation of two or more components from a liquid stream and/or concentration of components by removal of water from a liquid stream. However, in all cases, the primary role of the membrane is to act as a selective barrier. The separation performance of the membrane is function of its chemical composition, the characteristics of the feed stream, and of the interactions between components in the feed stream and the membrane surface (Lin, Rhee, and Koseoglu, 1997).

The ultrafiltration devices can be classified in two configurations based on the feed mass transfer characteristics of the membrane modules i.e. dead-end and tangential flow configurations. In dead-end ultrafiltration module, the feed stream is perpendicular to the membrane surface which results in the continuous accumulation of components on it, requiring periodic interruption of the process to clean or substitute the membrane (Figure 2).

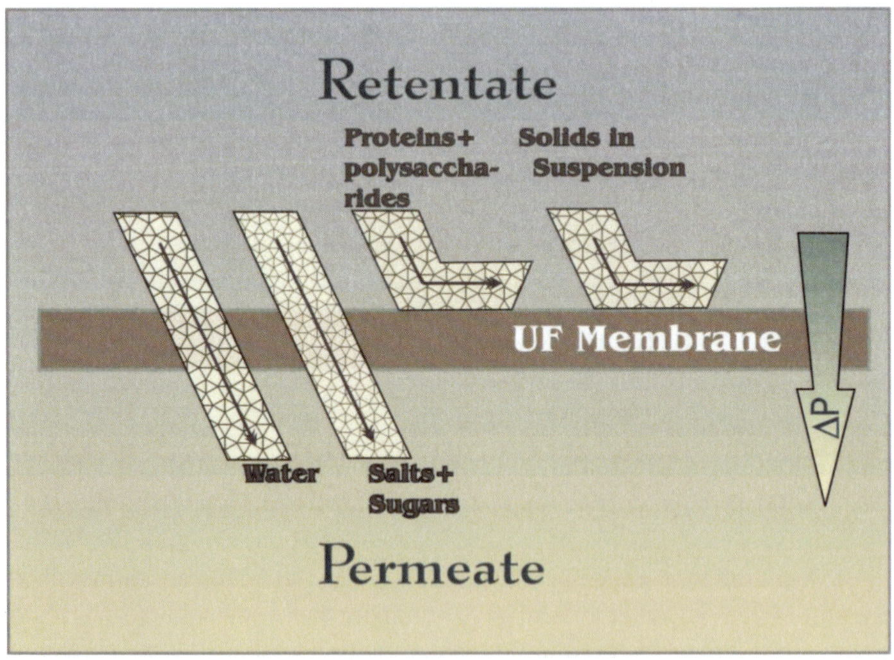

Figure 1. Ultrafiltration membrane process and its separation characteristics.

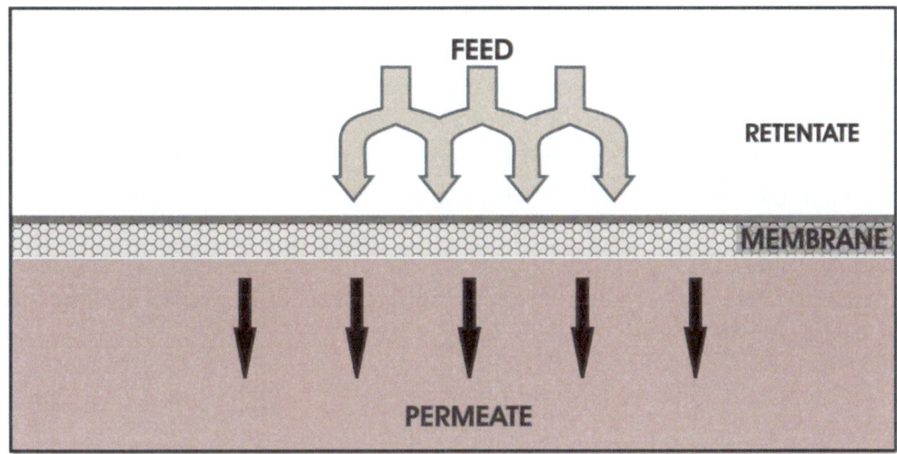

Figure 2. Dead-end ultrafiltration configuration.

However, most large-scale ultrafiltration modules are operated in tangential flow mode in which the feed stream is parallel to the membrane surface resulting in high shear rate and/or turbulence in the immediate viscinity of the membrane. The components that tend to accumulate on the surface are partly back-transported away from the membrane making the process more efficient than the dead-end configuration (figure 3).

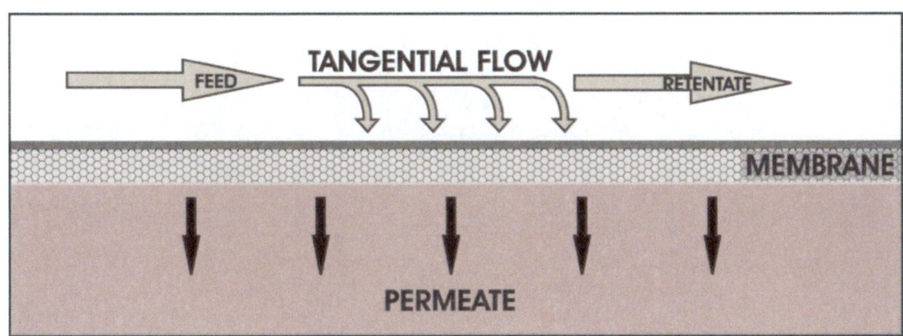

Figure 3. Tangential flow ultrafiltration configuration.

Advantages given by the use of this technology include: Separation of dissolved molecules assuming the appropriate membrane is used, product purification with concentration, the absence of a change in phase or state of the solvent during the process, and low equipment and operation (pumping) costs. Other advantages are that ultrafiltration can be operated at ambient temperature and that no complicated heat transfer or heat-generating equipments are needed since only electrical energy is required to drive the pump (Cheryan, 1998). For the chemical, pharmaceutical, biotechnological and food industries, ultrafiltration stand out as alternatives to conventional processes and in many cases the separation efficiency is greater and the final product quality is also improved (Strathmann, 1990). However there are also some limitations to ultrafiltration process such as membrane fouling which may result in low productivity, frequent cleaning of the membrane, and ultimately modification of membrane rejection properties and thus shortened membrane life. Also, ultrafiltration can not take the solutes to dryness (Cheryan, 1998).

3.1. Membrane Characteristics and Types

Most membranes used in ultrafiltration are classified as asymmetric membranes and are characterized by a thin skin on the surface of the membrane. The layers underneath the skin may consist of voids which serve to support the skin layer (Figure 4).

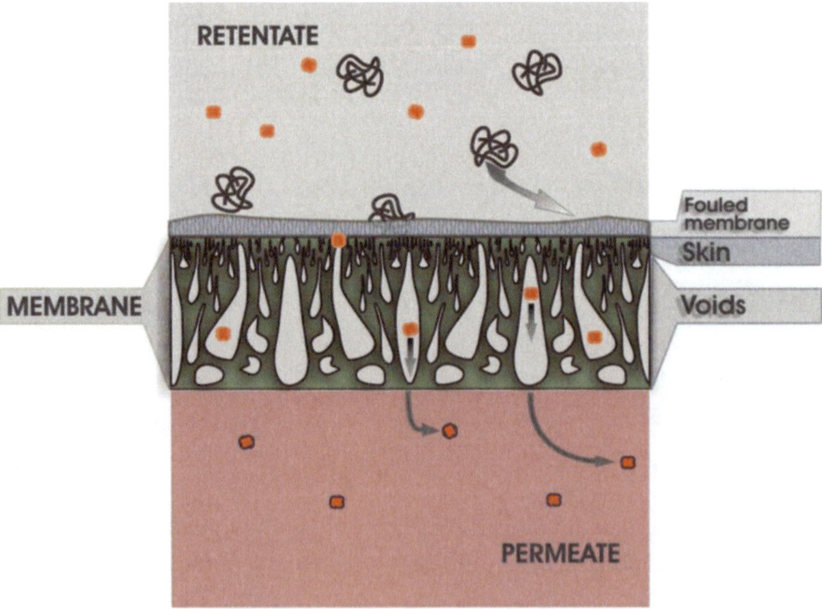

Figure 4. Schematic representation of the ultrastructure of an asymmetric ultrafiltration membrane.

Rejection of the components occurs at the skin layer and the retained components above the nominal molecular weight cut-off do not enter the main body of the membrane and consequently the membrane rarely get plugged. However, it is still subject to flux-lowering phenomena (Cheryan, 1998). By definition, the retention limits of ultrafiltration asymmetric membranes is expressed in terms of nominal rating, i.e. it refers to the molecular size or molecular weight above which a certain percentage of a specific component (with a specific molecular weight) in the feed stream will be retained by the membrane under controlled conditions (Cheryan, 1998). Commercial ultrafiltration membranes are produced either from polymers consisting of organic material such as: cellulose acetate, polyamide, polysulfone and polyvinylidene difluoride amongst others; or from inorganic materials such as: metals and ceramic materials (Cuperus and Nijhuis, 1993; Cheryan, 1998). Some advantages of inorganic membranes are their resistance to chemicals, their wide temperature and pH limits with some inorganic membranes that can be operated at temperature as high as 350 °C and at pH ranging between 0.5 to 13. Inorganic membranes are also able to tolerate frequent cleaning procedures and have consequently extended operating lifetimes (Cheryan, 1998). On the other hand, inorganic membrane are shock sensitive, i.e. they can be damaged if dropped or subjected to undue vibrations, a large pumping capacity is required to operate them at the recommended velocities of 2-6 m/sec, and the major limitation is their high purchase cost. Cheryan (1998) indicated that spiral-wound ultrafiltration membrane plants could be implemented for $225-$350/m^2 while a comparable ceramic system would cost $2200-$6000 /m^2 to put in place. For a more detailed review of membranes chemistry, structure and function, the readers are referred to the work of Zeman and Zydney (1996), Cheryan (1998) and de Morais Coutinho et al. (2009).

3.2. Fouling and Limitations of Uses

A typical permeate flux versus time curve is shown in Figure 5. The permeate flux versus time curve can be divided in three periods. An initial period, characterized by a rapid flux decrease attributed mainly to concentration polarization occurring very rapidly, a second period corresponding to a less severe decrease of the flux due to the effect of interactions between the material making up the membrane and the solute, a phenomenon known as "irreversible" fouling, and a third period corresponding to a small flux decrease until a pseudo steady-state is reached. The latter corresponds to particle deposition at the membrane surface, a phenomenon known as "reversible" fouling and due to the consolidation of irreversible fouling (Marshall and Daufin, 1995; de Morais Coutinho et al., 2009). In addition to the decay in permeate flux, these phenomena may also result in an alteration of the membrane selectivity i.e. that some components that would generally permeate the membrane can be retained by the cake formed at its surface. These changes eventually require extensive cleaning or replacement of the membrane (Zeman and Zydney, 1996).

In practice, the flux behaviour is characterized by the different hydraulic resistances. Evolution of the different hydraulic resistances is often described using Darcy's law (Eq. 1):

$$J = \frac{\Delta P}{\mu\, R_G}$$

(1)

where J is the permeate flux, ΔP is the transmembrane pressure, μ is the permeate viscosity and R_G is the global resistance. The global resistance is equal to the sum of the membrane resistance, the resistance due to concentration polarization, the resistance due to irreversible fouling and the resistance due to cake formation. The relative importance of the different resistances changes during ultrafiltration.

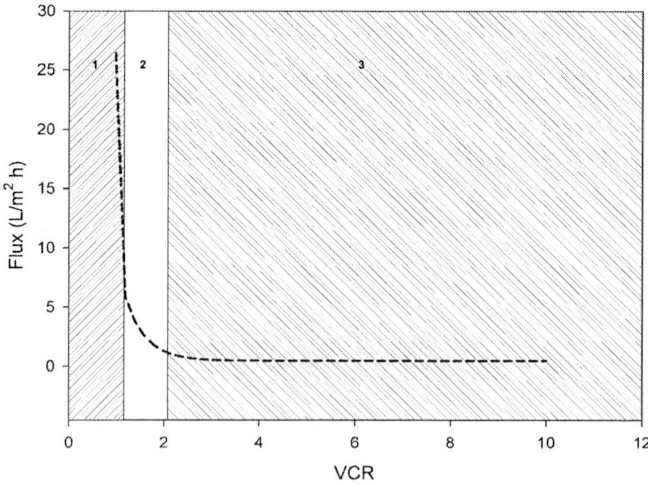

Figure 5. Typical flux versus volume concentration ratio (VCR) curve of ultrafiltration.

3.3. Operating Variables

The main operational parameters that affect the permeate flux are: pressure, temperature, feed concentration, and the tangential velocity (Scott, 2003). In theory, the permeate flux is directly proportional to the applied transmembrane pressure. However, in practice, this is rarely observed since an increase in the transmembrane pressure will also result in an increase in the global resistance due to more components brought at the membrane surface and in some cases due to compaction of the cake layer. Assuming there are no effects of temperature on the membrane fouling such as precipitation of components, or modification of the components structure, an increase in temperature will usually result in a decrease in the feed stream viscosity and an increase in diffusivity which will lead in higher permeate flux. An increase in feed concentration will result in an increase in feed stream viscosity and density, and higher concentration can also increase diffusivity. An increase in the feed density will have a negative impact on the permeate flux since for a given volume of solvent permeating the membrane, the amount of components brought at the membrane surface will be more important. However, an increase in diffusivity will have the inverse effect on the permeate flux since the back-transport of the components from the membrane surface to the bulk solution is improved. Thus, permeate flux may increase with feed concentration, decrease or not change. However, in practice, it is more common to observe a decrease in the permeate flux with an increase in the feed concentration. Finally, an increase in tangential velocity will generally increase the permeate flux due to the greater turbulence observed near the membrane surface. The greater turbulence results in an increase of the back-transport of components away from the membrane surface reducing the global resistance of the system (Strathmann, 1990; Cheryan, 1998). The module configuration (flat sheet, tubular modules, hollow fibers, plates units, spiral wound) also affects membrane performance.

3.4. Applications of Ultrafiltration in the Food Industry

It is beyond the scope of this chapter to review in details all the applications of ultrafiltration in the food industry. However, it worth to mention that ultrafiltration is used in the dairy industry to concentrate skim milk (concentrate the calcium and protein), for cheese manufacture including the recuperation of cheese whey proteins. It is also used in the sugar refining industry to remove some impurities from the thin juice prior to evaporation and crystallization and centrifugation to obtain white sugar and molasses. Ultrafiltration can also be used in almost all stages of vegetable oil production and purification (Cheryan, 1998; de Morais Coutinho et al., 2009) and it is also vastly applied in processing of fruit juices for their clarification. Ultrafiltration also finds applications for the isolation of animal and plant proteins. The ultrafiltration process has some significant advantages for the isolation of proteins when compared to the traditional isoelectric precipitation process.

3.5. Production of Pea Protein Isolate – Isoelectric Precipitation Process

Pea protein isolate and plant protein isolates, in general, are traditionally prepared by isoelectric precipitation process (Owusu-ansah and Curdy, 1991). The isoelectric precipitation process for pea protein isolate production consists in milling of the peas, solubilization of the proteins in 30–50°C water adjusted to pH 8 to 11 with a base, followed by centrifugation to remove the insoluble components. Starting material for the protein solubilization step can also be the pea residue resulting from the starch extraction process. The pea proteins present in the supernatant are then precipitated out at their isoelectric pH (4.2 to 4.5) by addition of a mineral acid, and they are recuperated by a second centrifugation step. The curd is suspended in water to remove the sugars and minerals and after reconcentration by applying a third centrifugation step, it is neutralized to pH 7 with a dilute base and dried using a spray-dryer to produce an isolate (Figure 6).

Yields of 60% of the original protein are reported in the literature for the isoelectric process (Owusu-ansah and Curdy, 1991). The isolates produced by isoelectric precipitation have poor solubility (Vose, 1980) possibly due to protein denaturation and to their high phytic acid content which alters the solubility of plant protein isolates especially at low pH (Omosaiye and Cheryan 1979; Brooks and Morr 1985). This process also requires large amount of water (extraction, washing of the curd and neutralization steps) and generates significant volume of effluents (isoelectric precipitation and washing steps) making it more or less attractive from an environmental point of view.

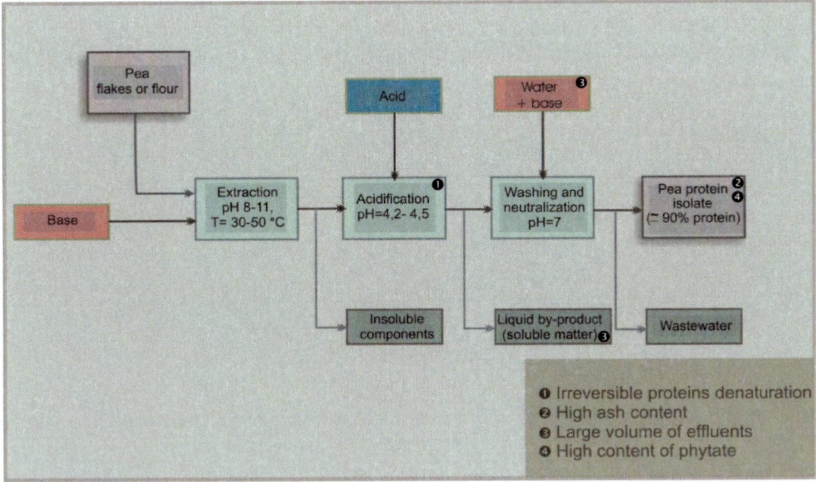

Figure 6. Isoelectric precipitation process for production of pea protein isolates.

In terms of composition, several factors can affect the composition of isolates produced by the isoelectric precipitation process. Those include pea composition, pea variety, pea maturity, as well as solubilizing and precipitation pHs. Concerning the latter, Gueguen (1983) found that isolates precipitated at pH 5.3 have higher protein content and lower lipid content than those precipitated at pH below 5.3.

3.6. Production of Pea Protein Isolate – Ultrafiltration Process

The ultrafiltration process begins with the solubilization of the proteins followed by centrifugation to remove insoluble components, similarly to the isoelectric precipitation process. However, the isoelectric precipitation step for isolation of the proteins and the washing step for the removal of some sugars and minerals are replaced by ultrafiltration-diafiltration, while the neutralization to pH 7 (if required) and spray-drying steps are also not modified. More specifically, the extract is fed to an ultrafiltration module and the permeate passing through the membrane, which contains minerals, sugars as well as some unwanted components is taken off. The retentate, which is enriched in protein, may be spray dried or further purified by diafiltration before being spray dried (Figure 7).

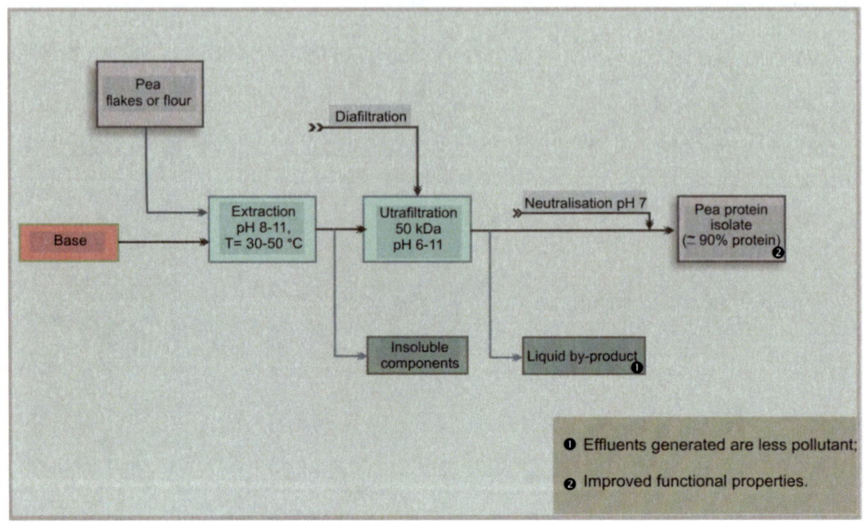

Figure 7. Isoelectric precipitation process for production of pea protein isolates.

Diafiltration can be performed in continuous or in discontinuous mode. Continuous diafiltration consists in the removal of permeable solutes by addition of water at the same rate as the permeate flux, thus keeping feed volume constant during processing, while discontinuous diafiltration consists in the removal of permeable solutes by redilution with water followed by re-ultrafiltration (Cheryan, 1998). One of the main advantages of the ultrafiltration process is that the volume of water required in the process is less than for the traditional isoelectric precipitation process since there is no need to resuspend the proteins in a dilute base prior to the spray drying since there is no proteins precipitation involved in the current approach.

Fuhrmeister and Meuser (2003) have investigated the impact of the molecular weight cut-off (5, 10, 30, 50 and 100 kDa) on the ultrafiltration process performance in terms of permeate flux and protein retention. Ultrafiltration of the extracts was performed with miniature plate modules with polyethersulfon membranes with 0.005 m^2 (Sartorius, Germany). The pea protein extracts to be processed by ultrafiltration were obtained by means of the TUB process for extracting starch from wrinkled peas developed by Meuser et al. (1997) except that the soluble components of the pulp were separated from the insoluble components by means of a beaker centrifuge (Varifuge 3.2RS, Heraeus, 20 min, 3000 g) after an extraction time of 1 h, and the supernatant was poured off and sieved with a 63 μm screen to remove the remaining fine fibres. For all experiments, the feed stream flow rate was set to

1.05 m/sec and a low transmembrane pressure (1 bar) was selected to minimize the membrane fouling. The permeate flux was measured until a concentration factor of 5 was reached. The temperature of the feed stream was set at 25 °C. As expected, the permeate flux increases with an increase in the membrane molecular weight cut-off. It rose from 5.5 to 17.1 l/m^2.h when the cut-off was increased from 5 to 50 kDa. By contrast, the permeate flux increased by only 1.5 l/m^2.h when the cut-off was raised from 50 to 100 kDa. Concerning the protein rejection, it slightly decreased from 80 to 75% when the cut-off was increased from 5 to 50 kDa and by an additional 3% when the cut-off was raised from 50 to 100 kDa. Thus, Fuhrmeister and Meuser (2003) conclude that the most favourable combination of permeate flux and protein rejection in economic terms was achieved using a 50 kDa hollow fiber membrane module.

Mondor et al. (2009a) reported the use of a laboratory-scale ultrafiltration system (Amicon, Model DC10L) equipped with a 50 kDa hollow-fibre membrane for the production of pea protein concentrate. They have considered three 50 kDa membrane with different fibre lumen internal diameter (0.5, 1.1 and 1.5 mm) (HF2-20-PM50; HF1-43-PM50 and HF1-60-PM50; Chalinox, St-Hyacinthe, Quebec). They found that for similar operating conditions (2% pea protein extract, pH 9, room temperature, transmembrane pressure of 15 PSI and feed flow rate of 0.13 l/sec), the importance of fouling was increasing with a decrease in the fibre lumen diameter. However, despite the high degree of fouling observed for the 0.5 mm module, it was for this module that the highest mean permeate flux was observed. These results were explained in terms of the shear rate within the fibres as well as based on the intrinsic membrane resistance. Since the fibre internal diameter was different for the three modules, this resulted in different shear rates within the fibres i.e. 35450 s^{-1} in the 0.5 mm module, 14650 s^{-1} in the 1.1 mm module, and 7880 s^{-1} in the 1.5 mm module. It is well known that an increase of the shear rate will result in an increase of the particles back-transport away from the membrane surface and consequently minimize cake formation (Cheryan, 1998; Kramadhati et al., 2002). Thus, it was expected to observe the least reversible fouling for the 0.5 mm module, yet here it was the contrary. However, it was also observed that the clean membrane resistance was different among the three modules with 1.40×10^{12} m^{-1}, 1.69×10^{12} m^{-1} and 3.49×10^{12} m^{-1} for the 0.5, 1.1 and 1.5 mm diameter module, respectively. The three modules having the same molecular weight cut-off, it was assumed that the difference in resistance was due to a difference in fibre thickness. A combination of a thin fibre and a small diameter would yield a fibre of sufficient solidity, but a larger diameter would require a larger thickness to ensure that the fibre is solid enough. Consequently, the 0.5 mm module which had the lowest membrane resistance had the highest initial flux ($1.99X10^{-5}$ m/s), the 1.5 mm module which had the highest resistance had the lowest initial flux ($1.22X10^{-5}$ m/s), while the 1.1 mm module, which had an intermediate resistance had an initial flux that was also intermediate ($1.59X10^{-5}$ m/s). As previously reported by Stamatakis and Tien (1993), Belfort et al. (1994) and Kramadhati et al. (2002), the force responsible for particle deposition (drag force) will increase with an increase in the permeate flux. As a result if the drag force is predominant over the back-transport of particles away from the membrane surface, we would expect the most significant cake formation for the 0.5 mm diameter module, followed by the 1.1 mm diameter module and by the 1.5 mm diameter module. No difference in concentrate composition was observed; the protein, mineral and phytic acid (estimated as total phosphorous) compositions were identical.

As it is the case for isolates produced by isoelectric precipitation, the composition of the isolates produced by ultrafiltration will also be affected by the pea composition, pea variety and pea maturity. In addition, it will be affected by the extraction conditions and the ultrafiltration and diafiltration pHs. Taherian et al. (2011a) reported the use of a home-made module equipped with two 50 kDa hollow fibres membranes with a surface of 2.3 m^2 for each membrane (Romicon, model CTG 3, HF25-60-PM50; KOCH Membrane Systems, Inc.,MA) for the production of pea protein isolates. The extracts to be processed by ultrafiltration-diafiltration were obtained by extraction in either water or 0.06 M KCl solution, at pH 7.5 and 25 ºC with a ratio of 1:15 (w/w), i.e. 7 kg flour in 105 kg extraction solution. The ultrafiltration step was carried out through the application of a volume concentration ratio (VCR) of 5 at pH of 7.5. Further purification was obtained by the application of discontinuous diafiltration (DF) step consisting of the adjustment of solution volume to its initial volume with tap water and re-VCR 5 at either pH of 7.5 or 6. Following these membranes treatments, the isolates were freeze-dried, placed in aluminium pouches, sealed, and stored at 4 °C prior to further examination. Their results indicated a protein content (N X 6.25: % dry basis) of 97.76±0.20, 90.73±0.01, 94.26±0.11 and 94.39±0.01, for the isolate extracted in H2O/pH 7.5/25 °C and purified by ultrafiltration at pH 7.5 and by diafiltration at pH 6, in H2O/pH 7.5/25 °C and purified by ultrafiltration at pH 7.5 and by diafiltration at pH 7.5, in 0.06 M KCl/pH 7.5/25 °C and purified by ultrafiltration at pH 7.5 and by diafiltration at pH 6, and in 0.06 M KCl/pH 7.5/25 °C and purified by ultrafiltration at pH 7.5 and by diafiltration at pH 7.5, respectively. The corresponding ash content was

2.43±0.02, 3.95±0.12, 4.77±0.16 and 5.27±0.02. Those results indicated that the isolate extracted in H2O/pH 7.5/25 °C and purified by ultrafiltration at pH 7.5 and by diafiltration at pH 6 has the highest protein content and the lowest ash content.

3.7. Effectiveness of Membrane Processing on Removal of Antinutritional Factors - Phytic Acid

Taherian et al. (2011a) have also investigated the removal of phytic acid using 50 kDa hollow fibres membranes for the production of pea protein isolates. Phytic acid content was estimated based on the Wade Reagent method reported by Gao et al. (2007) with slight modification (Taherian et al., 2011a). Results indicated that isolates purified by diafiltration at pH 6.0 had lower phytic acid content than the ones purified at pH 7.5 (2.45 vs 4.66 mg phosphorus from phytic acid (PA-P)/g protein for water extracts; 2.84 vs 5.57 mg phosphorus from phytic acid (PA-P)/g protein for KCl extracts). For all membrane purified pea protein isolates the amount of phytic acid was less than the one reported for commercial isolate (7.71 mg phosphorus from phytic acid (PA-P)/g protein). Similar conclusions were draw when the total phosphorous content was used as an estimate of the phytic acid content. These results are in agreement with previous observations reported by Mondor et al. (2009b) for the impact of ultrafiltration/diafiltration process pH on the removal of phytic acid during the production of full fat and defatted chickpea protein concentrates, as well as with the results reported for the production of soy protein isolates by membrane technologies (Omosaiye and Cheryan 1979; Mondor et al., 2004; Ali et al., 2010).

3.8. Other Antinutritional Factors

The literature dealing with the impact of membrane processing on the antinutritional factors content, other than phytic acid content, of pea protein isolate is scarce. However, Frederickson et al. (2001) reported that production of pea protein isolates using a 50 kDa plate frame ultrafiltration membrane enabled to reduce the level of oligosaccharides from 176 ± 8 mg/g in the protein extract to 12.1 ± 0.9 mg/g in the ultrafiltration retentate, when an ultrafiltration concentration factor of 3 was applied followed by a diafiltration factor of 3. When compared to the oligosaccharides content of the starting pea seeds 77.3 ± 3.7 mg/g, the amount of oligosaccharides in the final isolate after drying is only at 11.7 ± 1.9 mg/g. This is a good indication that ultrafiltration using a 50 kDa membrane is efficient to remove a significant fraction of the pea oligosaccharides. For the production of chickpea protein concentrates, using a 50 kDa hollow fiber membrane module (Quixstand module from GE Healthcare, Ste-Anne de Bellevue, Qc, Canada) and applying a volume concentration ratio of 5 during the ultrafiltration step and a continuous diafiltration step until 4 volumes of water were processed, Mondor et al. (2009b) reported that for most conditions considered in their work the phenolic content of protein concentrates produced by ultrafiltration-diafiltration had a higher total phenolic content than their counterpart produced by isoelectric precipitation. It was also reported that the concentrates produced by diafiltration at pH 6 had higher total phenolic content than their counterpart produced by diafiltration at pH 9. For the trypsin inhibitor content, the same authors reported that for most conditions no significant difference was observed in the trypsin inhibitor content between the protein concentrates produced by ultrafiltration-diafiltration versus the ones produced by isoelectric precipitation. It was also reported that for the ultrafiltration-diafiltration process, the protein concentrates produced from the full fat flours by ultrafiltration pH 9/diafiltration pH 6, generally had higher trypsin inhibitor content than their counterparts produced by ultrafiltration pH 9/diafiltration pH 9. These results were explained in terms of the more severe cake formation observed during the diafiltration processing of full fat extracts at pH 6. The more severe cake formation observed at that pH would have resulted in a higher retention of the trypsin inhibitors, than for the experiments carried out at pH 9. While not confirmed for the production of pea protein isolates, theses results suggest that both polyphenols and trypsin inhibitors are not efficiently removed by ultrafiltration with a 50 kDa membrane.

3.9. Importance of Pre-Treatment of Materials Prior to Membrane Processing

As aforementioned, while membrane processing can be helpful to produce pea protein isolates with low level of phytic acid and oligosaccharides, it may not significantly reduce the level of other antinutritional factors such as trypsin inhibitors and polyphenols. Consequently, pre-treatment of the pea seeds/flours prior to membrane processing is necessary to obtain isolate with low level of other antinutritional factors and can also be helpful to further decrease the level of phytic acid in the isolate. Among the different pre-treatments, soaking, dehulling, germination and heat treatments (cooking, roasting, autoclaving and extrusion cooking) of the seeds/flours are the most applied. However, when applying these pre-treatments, the users must also take into consideration their impact on other constituents such as storage proteins. For example, during germination proteases are activated and they will be responsible for proteins hydrolysis, while some severe heat treatments could result in proteins denaturation. The different pre-treatments that can be applied to reduce the level of antinutritional factors are reviewed below for each antinutritional factors.

3.9.1. Trypsin Inhibitors

Ma et al. (2011) observed that both roasting of the pea flour (1 min in an oven preheated at 80°C) and boiling in a water bath at 90 °C for 20 min (after hydration in Millipore water under agitation for 1 h at 20 °C, 10% w/v) resulted in significant reductions in the trypsin inhibitor activities of the pea flour (approximately by 40%). However, the effect of roasting did not differ significantly from that of boiling. These results are in agreement with the findings of Jourdan et al. (2007), who reported that the trypsin inhibitor activities of common bean were reduced by between 80% and 95% after soaking and cooking in a 90 °C water bath for 15 min and were completely eliminated after cooking at 90 °C for 40 min, and with the results of Wang et al. (2003) who reported that cooking of yellow field pea for 30 min reduced in average the trypsin inhibitor activity by 84.3%. In another study, Wang et al. (2008) investigated the impact of soaking, cooking and dehulling on six pea varieties (Nitouche, Keoma, SW Parade, Elipse, Delta and CDC Mozart) from producers across western Canada. An increase in trypsin inhibitor activity by 3.2-19.3% was observed when the peas were soaked in distilled water for 24 hours (ratio of 1:4 seed:water, w/w at room temperature). This was attributed to the higher retention of trypsin inhibitors in soaked seeds. However, these results differ from the ones of Roman et al. (1987) who reported a decrease of 28% in the trypsin inhibitor activity for soaked cowpeas and to the ones of Abd El-Hady et al. (2003) who reported a decrease of 15.4% in the trypsin inhibitor activity for soaked peas. Wang et al. (2008) reported a significant reduction in the trypsin inhibitor activity (62.3–77.8%) for all samples after soaking in distilled water at a ratio 1:4 (seed:water, w/w) for 24 h at room temperature followed by cooking in a boiling water bath for its predetermined cooking time, as described by Wang and Daun (2005). Finally, dehulling resulted in a significant reduction in trypsin inhibitor activity (5.3–13.1%) for the six pea varieties considered. In another study, Van der Poel et al. (1992) observed that extrusion cooking at temperature ranging between 106 and 140 °C while the moisture content was ranging between 14.7 and 32.7% was very efficient to reduce the trypsin inhibitor activity in round-seeded pea (var Finale) but was less efficient in wrinkle-seeded peas (var C306). For the round-seeded peas, the residual trypsin inhibitor activity was always less than 0.10 mg/g for the extruded peas as compared to 1.75 mg/g in the raw seeds, while it vary between 0.34 and 1.96 mg/g for the extruded wrinkle-seeded peas, as compared to 3.56 mg/g in the raw seeds. In the same context, Abd El-Hady et al. (2003) reported a complete inhibition of trypsin inhibitor in peas by extrusion at either 140 °C or 180 °C and at moisture content of 18 or 22%. Thus, based on the aforementioned studies, it appears that the heat treatments are the most efficient processes to reduce the trypsin inhibitor activity. In another study, Alonso et al. (1998) observed that for three pea cultivars (cv. Renata, Solara and Ballet) germination at 25°C for 24, 48 or 72 hours significantly reduced the trypsin inhibitor activity of the pea seeds when compared to the ones of the raw seeds (except for the Renata germinated for 24 hours). It is the germination at 25 °C for 72 hours which was the most efficient to reduce the trypsin inhibitor activity of the seeds (from 3.80±0.24 to 2.76±0.11 IU /mg dry basis for Renata; from 2.80±0.09 to 0.70±0.03 IU /mg dry basis for Solara; from 6.32±0.23 to 1.55±0.09 IU /mg dry basis for Ballet). These results are in agreement with the ones of Bishnoi et al. (1994) who reported that for four pea cultivars (cv. Bonneville, Arkel, HFP4 and Rachna) after 48 hours germination, pea seeds had the minimum trypsin inhibitor activity with 53-59 % reduction as compared to that in raw seeds.

3.9.2. Polyphenols

Alonso et al. (1998) observed that for three pea cultivars (cv. Renata, Solara and Ballet) both soaking and dehulling of the pea seeds had no significant effect (p<0.05) on their polyphenol content (mg/100 g dry basis). A possible explanation is that most of the polyphenols in Renata, Solara and Ballet are not located in the seed-coat. However, these results differ from the ones of Bishnoi et al. (1994) who reported that soaking of pea seeds (HFP4 and Rachna) in double distilled water for 6, 12 and 18 hours at 30 °C in an incubator and seed to water ratio of 1:5 (w/v) could reduce the content of polyphenols in pea seeds by as much as 58%. On the other hands, Alonso et al. (1998) reported that extrusion cooking at 148°C while maintaining the moisture content at 25% showed to significantly decrease the level of polyphenols for all three cultivars (from 37±4 to 25±1 mg/100 g dry basis for

Renata; from 39±5 to 23±2 mg/100 g dry basis for Solara; from 50±9 to 23±3 mg/100 g dry basis for Ballet). However, it is the germination at 25 °C for 72 hours which was the most effective treatment to reduce the polyphenols content of the seeds (from 37±4 to 19±3 mg/100 g dry basis for Renata; from 39±5 to 12±2 mg/100 g dry basis for Solara; from 50±9 to 13±1 mg/100 g dry basis for Ballet). These results are in agreement with the ones of Bishnoi et al. (1994) who reported that after 48 hours germination, pea seeds had the minimum content of polyphenols with 84-88 % loss as compared to that in raw seeds. In another study, Abd El-Hady et al. (2003) reported that both soaking (in distilled water (1:5, w/v) at 30°C for 16 hours) and extrusion at either 140 °C or 180 °C reduced significantly the level of polyphenols in pea seeds. The most significant decrease in the polyphenols content was observed for soaked pea seeds extruded at 180 °C with a moisture content of 22% (from 460 mg/100 g for raw seeds down to 393 mg/100 g for soaked seeds and to 343 mg/100 g for soaked and extruded seeds).

3.9.3. α-Amylase Inhibitor

Alonso et al. (1998) reported that after dehulling, the α-amylase inhibitor activity of dry pea seeds (cv. Solara) was increased from 16.8±1.55 IU/g (dry basis) in the raw seeds to 20.3±0.39 IU/g (dry basis) in the dehulled seeds. This increase might be attributed to the higher concentration of this inhibitor in the cotyledons fractions than in the seed coats. The same authors reported that soaking in the dark in double-deionized water (1:5 w/v) for 12 h at 30 °C followed by drying the seeds at 50 °C reduced the α-amylase inhibitor activity down to 12.2±0.84 IU/g (dry basis), while a reduction of 48% in the α-amylase inhibitor activity was observed when the Solara pea seeds were germinated for 72 hours. However, it is the extrusion cooking (148°C and 25% moisture) which was the most effective process to reduce α-amylase inhibitor activity since the amylase inhibitor of the Solara cultivar became completely inactive when seeds were extruded using the aforementioned conditions. No α-amylase inhibitory activity was observed in the raw seeds for the (cv. Renata and Ballet). This agrees with the results obtained by Jaffé et al. (1973) in Pisum sativum where, from the seven varieties studied, only five had α-amylase inhibitory activity. Concerning the reduction in activity on exposure to heat treatment, as reported by Shekib et al. (1988), the reduction in activity varied enormously in legume seeds.

3.9.4. Lectins

Literature dealing with the removal of lectins from pea seeds by various pre-treatments is scarce. This is possibly due to the fact that the lectins content of pea seeds is low when compared to the one of beans (Campos-vega et al., 2010). However, in their work, Alonso et al. (1998) reported that only extrusion cooking had a significant effect on the haemagglutining activity of pea seeds with a reduction of 98% in Renata and Ballet (from 5100 HU/g dry matter down to 100 HU/g dry matter for Renata; from 6000 HU/g dry matter down to 100 HU/g dry matter for Ballet), and a complete elimination in Solara (from 6200 HU/g dry matter down to 0 HU/g dry matter for Solara). Soaking, dehulling and germination of the pea seeds for 24, 48 or 72 h at 25 °C did not significantly change the haemagglutinating activity of the seeds. These results differ with the ones of El-Adawy et al. (2003) who reported a significant decrease in the haemagglutinating activity of flour produced from pea seeds (variety Lencolen) germinated for 72 and 120 h, in the dark and at room temperature, after soaking in tap water for 19 h at room temperature (seed to solvent ratio 3:10, w/v). The haemagglutinating activity decreased from 6400±30 HU/g flour produced from raw seeds down to 1600±10 HU/g flour produced from seeds germinated for 72 h and to 1400±10 HU/g flour produced seeds germinated for 120 h.

3.9.5. Saponin

As reported by Campos-vega et al. (2010), the saponin content of pea seeds is low being in the order of 0.1-0.3 mg/100 g. This may explain the low number of publications related to the removal of saponin from pea seeds by various pre-treatments. Shi et al. (2004) claimed that, even if saponins have long been considered undesirable due to their haemolytic activity and toxicity, there is enormous structural diversity within this chemical class, and only a few are toxic. In their work, Bishnoi and Khetarpaul (1994) have characterized the saponin content of vegetable peas (Bonneville and Arkel) and field peas (HFP4 and Rachna) submitted to various treatments including, soaking, dehulling, ordinary cooking, pressure cooking and germination. Soaking treatment consisted in suspending the peas in double distilled water for 6, 12 or 18 hours at 30°C in an incubator, with a seed to water ratio of 1:5 (w/v). For the ordinary cooking, the soaked seeds (12 hours) were cooked on a hot plate until they became soft using a seed to water ratio of 1:3 (w/w). Unsoaked seeds were also cooked similarly, using seed to water ratio of 1:4 (w/v). Both soaked (12 hours) and unsoaked seeds were pressure cooked at 121 °C for 10 min. The dry seeds to cooking water ratio was 1:3 (w/v) for unsoaked seeds, whereas it was 1:2 (w/v) for soaked seeds. For germination, the soaked seeds (12 hours) were placed in sterile petri plates, lined with wet filter papers, and kept in an incubator at 30 °C for 12, 24 and 48 hours. All the treated samples were dried to a constant weight in oven at 60°C. Saponin contents of raw pea seeds were in the amount of 251, 240, 119 and 109 mg/100 g for the Bonneville, Arkel, Rachna and HFP4 varieties, respectively. Soaking significantly reduced the saponin contents of pea seeds, reduction increasing with an increase in soaking time (186.5, 182.4, 84.4 and 80.6 mg/100 g for the Bonneville, Arkel, Rachna and HFP4 varieties after 18 hours soaking, respectively). Dehulling of the soaked seeds further brought significant losses of saponins (P<0.05) in the order of 4.1, 4.1, 11.6 and 9.1%, when compared to the saponin contents of the raw seeds, for the Bonneville, Arkel, Rachna and HFP4 varieties, respectively. Ordinary and pressure cooking significantly reduced the saponin contents of unsoaked, soaked and soaked-dehulled pea cultivars. Pressure cooking of the soaked and soaked-dehulled peas showed to be the most efficient treatment with a reduction in the saponin contents up to 50-68% in different cultivars. Finally, germination of pea seeds for different periods reduced significantly their saponin content. Germination of peas for 48 hours showed the maximum saponin reduction with final saponin contents of 80.1, 78.8, 28.5 and 17.2 mg/100 g for the Bonneville, Arkel, Rachna and HFP4 varieties, respectively.

3.9.6. Phytic Acid

As previously mentioned, under some specific conditions, ultrafiltration is relatively efficient to remove phytic acid from pea protein extracts, in order to obtain pea protein isolates with low level of phytic acid, when compared to isolates produced by the traditional isoelectric precipitation process. However, it seems logical that the combination of ultrafiltration with some of the aforementioned pre-treatments could further decrease the level of phytic acid in the final pea protein isolates. Studies carried-out by Bishnoi et al. (1994), Alonso et al. (1998), El-Adawy et al. (2003), Abd El-Hady and Habiba (2003), Wang et al. (2008) indicated that soaking of the pea seeds will under most conditions significantly reduce its phytic acid content (by as much as 12.0%), heat treatments (by as much as 16.0%) such as ordinary cooking, pressure cooking and extrusion cooking, as well as germination (by as much as 81.7%) also significantly reduce the phytic acid content of pea seeds. On the other hands, dehulling of the pea seeds results in an increase of its phytic acid content as reported by Alonso et al. (1998) (by as much as 13.7%) and by Wang et al. (2008) (by as much as 8.0%).

However, understanding the limitations caused by these antinutional factors for pea proteins to act as a functional ingredient is extremely important. Following, after a brief introduction to functional properties of legume proteins, attempts to remove or decrease these limiting factors via membrane processing and a comparative study on functional properties of commercial and membrane processed pea protein isolate is provided.

4. FUNCTIONAL PROPERTIES OF LEGUMES PROTEINS

Functionality is defined as any property of a food or food ingredient that affects its utilization except its nutritional ones. Consequently, functional properties of plant proteins in either form of concentrate or isolate govern the extent of its utilization in food products. Numerous studies have shown that the development of new food products depends upon knowledge of functional properties of individual protein as an important step towards their evaluation and utilisation (Owusu-Apenten, 2002; Tsumura et al. 2005; Wang et al., 2010; Nunes, Raymundo, Sousac 2006; Liu et al. 2010; Makri, Papalamprou, Doxastakis, 2005; Mwasaru et al., 2000). For instance, plant proteins in form of isolate have widely been employed to improve the texture and generate gel structures that give the same texture with reduced lipid in the final product as well as to encapsulate the bioactive ingredients (Shand et al., 2007; Taherian et al., 2011a). Whole pea flour derived from yellow field peas contributed desirable functional characteristics to a wide range of food products due to its high levels of water and oil absorption, good gelation capabilities, and gel clarity (Agboola et al., 2010).

The chief functional properties that are evaluated for development of new product are solubility or the degree to which protein can be dissolved, rheological properties, emulsification, gelling and foaming abilities. These properties are intrinsic physicochemical characteristics, which affect the behaviour of proteins in food system during processing, manufacturing, storage and preparation. Consequently, the technological uses of pea proteins depend largely on the physicochemical properties which are necessary for their successful incorporation into food systems (Kinsella, 1979; Adebiyi and Aluko, 2011).

The differences in functionality of legumes proteins have also been related to the protein compounds (Makri, Papalamprou, Doxastakis, 2005). For example the pea proteins, have been reported to form heat-induced gels with purified vicilin, but not legumin. The broad bean protein isolate required the lowest concentration to form a gel, meanwhile the isolate obtained from the pea seeds needed the highest concentration.

While the protein solubility of the protein isolates of pea, broad bean and soybean revealed similar patterns, the water and oil absorption capacities of broad bean protein isolate had the highest values. The soybean had the poorest capacity regarding these functional properties but its foam expansion and foam ability is higher and close to that of the pea protein isolate respectively (Makri, Papalamprou, Doxastakis, 2005).

Among the aforementioned functional properties, solubility profile is an excellent indication of protein functionality. Good solubility can markedly expand potential applications of a protein. In general, all other functional properties are related to the aqueous solubility of proteins. The positive correlation between solubility and the ability of a protein to function as emulsifier, gelling agent, and viscosity builder has been reported in many studies (Boye et al., 2010; Taherian et al., 2011a).

The protein hydrophilicity/hydrophobicity is also considered to play important roles in the functional properties of food proteins. Surface hydrophilicity/hydrophobicity and interfacial tension are interrelated and both have shown significant correlations with emulsifying activity of the proteins. It has also been suggested that the emulsification of oil with protein can be explained using the concept of protein hydrophilicity/hydrophobicity (Voutsinas, Cheung, and Nakai, 1983). However, factors such as pH, ionic strength, and the presence of antinutritional components will affect the functional properties of pea protein.

4.1. Membrane Processing and Enhancement of Functional Properties of Pea Proteins

While many factors influence the performance of pea proteins in food systems, protein structure (native versus denatured) and antinutritional factors content are the most important factors affecting the solubility and, hence, the other functional properties of pea proteins. Attempt to keep the proteins in their native state and to remove these factors and purify the extracted proteins via membrane processing have been subjects of recent studies. Boye et al. (2010) compared the functional properties of pea, chickpea and lentil protein concentrates processed using ultrafiltration and isoelectric precipitation techniques. For isoelectric precipitation of pea proteins, the optimal extraction conditions were pH 9.5, 1/15 solid/liquid ratio and 35°C.

For ultrafiltration, a 50 kDa MWCO membrane was used and the extracts were concentrated by applying a volume concentration ratio of 5 prior to diafiltration (4X) at pH 6.0. The initial protein content of the pulses (16.7–24.8%, w/w) was concentrated nearly 4-fold. They claimed that a slightly higher protein contents (69.1–88.6%, w/w) was obtained using ultrafiltration process, while the isoelectric precipitation process resulted in 63.9–81.7%, w/w protein. The yellow pea concentrate purified by ultrafiltration indicated higher solubility at pH 3.5 and higher fat adsorption capacity than its counterpart produced by isolectric precipitation, in addition, the protein concentrate produced using ultrafiltration had slightly better gelling properties. They also mentioned that all the isolated proteins showed comparable solubility, fat binding capacity, emulsifying stability and foaming properties to those reported for soy protein isolate.

Comparative study on solubility, emulsion stability, rheological and mechanical properties of solutions and heat-induced gels of commercial and membrane processed by Taherian et al. (2011a) concluded that under some specific conditions (i.e. diafiltration at pH 6), membrane processing is capable of lowering down the phytic acid content and improves solubility of the pea protein. They also claimed that properties such as emulsion stability, rheological and mechanical properties of solutions, foaming stability and heat-induced gels for membrane processed pea protein isolates were superior compare to those of commercial isolate. They related the lower solubility of commercial pea protein isolates to protein denaturation and to the formation of insoluble complexes at acidic pH causing the reduction of emulsifying capacity pH 3.4. They also claimed that membrane purification could provide new opportunities to extend the range of functional properties of pea protein isolates in different food systems.

Impact of processing on functional properties of protein products from wrinkled peas was also investigated by Fuhrmeister and Meuser (2003). They found that the membrane processed wrinkled pea protein were superior to those obtained by isoelectric precipitation, in particular with regard to their solubility (96.7% vs 56.6% at pH 7) but also in terms of their fat binding capacity, their emulsifying activity and their foam-forming capacity. In this respect the products could be compared to commercially produced protein isolates from soybeans and smooth peas. In some cases they were even superior to these.

5. EMULSIFICATION OF OMEGA-3 FISH VIA MEMBRANE PROCESSED AND COMMERCIAL PEA PROTEIN ISOLATES IN PRESENCE OF MALTODEXTRIN: A COMPARATIVE STUDY

In the formulation and development of traditional and novel foods, emulsification is one of the most important functionality of proteins and amphoteric hydrocolloids. Several formulated foods in the market are in form of emulsions and proteins with good surface properties and solubility are desirable as emulsifying agents (Jianmei et al. 2007; Tsumura et al. 2005; Bouaouina et al. 2006; Yufei et al. 2005; Taherian et al., 2011b).

The emulsifying property of protein depend not only on factors connected with the preparations themselves, such as kind of protein, method of preparation and the way by which it has been produced, but are also influenced by environmental factors. These factors may explain the differences in the emulsifying properties in various protein products.

Our previous study showed that solubility of yellow pea protein largely depends upon its isolation process (Boye et al., 2010; Taherian et al., 2011a) and experimental ultrafiltration conditions. The investigation showed that while commercial pea protein isolate exhibit similar pattern, its solubility was significantly lower that that of membrane processed pea protein isolates (Taherian et al., 2011a). The differences in protein solubility–pH relationships were related to the dissimilarity of protein constituent composition and surface hydrophobicity. It was also suggested that the lower solubility of commercial pea protein isolate may have arisen from an increase in exposed hydrophobic residues, leading to increased hydrophobic interaction between proteins and/or peptides in the acidic pH region, as well as due to its higher phytic acid content.

The following investigation was conducted with a particular concern on comparing encapsulation capacities of commercial and membrane processed pea protein isolates. With this regard, we examined the rheological and emulsifying properties of membrane processed or commercial pea proteins stabilized fish oil-in-water emulsions in conjugation with maltodextrin at the constant pH of 7 using deposing techniques. The aim of this study is to augment the stability of dispersed fish oil in order to reduce its susceptibility to physical separation.

6. MATERIALS AND METHODS

6.1. Materials

Fish oil (OmegaPure, Houston, TX) containing 32-37% omega-3 fatty acid was provided by NEX-XUS (Montreal, PQ). Based on the claim by OmegaPure the fish oil contains 35.2% omega-3 fatty acids and fatty acid profile was as follow:

Fatty acid	Area% of total fatty acid
Myristic, C14:0	8.2
Palmitic, C16:0	19.1
Palmitoleic, C16:1	11.7
Stearic, C18:0	3.0
Oleic, C18:1	13.2
Linoleic, C18:2 (n-6)	2.2
Alpha Linoleic, C18:3 (n3)	1.6
Stearidonic, C18:4 (n-3)	3.5
Arachidonic, C20:4 (n-3)	1.7
Eicosapentaenoic, C20:5 (n-3)	13.8
Docosapentaenoic, C22:5 (n-3)	2.2
Docosahexaenoic, C22:6 (n-3)	11.8
Other	7.0

Commercial pea protein isolate (Propulse™ Pea Protein) was purchased from Nutri-Pea Ltd (Portage La Prairie, MB). Based on the information provided by Nutri-Pea the isolate contains 82±2% protein (Dumas, N × 6.25), 11000 ppm phosphorus, <4.0% ash and <3% fat.

Membrane processed pea protein isolates were made in pilot plant using water or KCl as extraction solutions at 25°C and pH of 7.5. The resultant solution was then purified/concentrated by ultrafiltration/diafiltration using an home-made module equipped with two 50 kDa hollow fibres membranes with a surface of 2.3 m^2 for each membrane (Romicon, model CTG 3", HF25-60-PM50; KOCH Membrane Systems, Inc., MA). Ultrafiltration (UF) step was first carried out through application of volume concentration ratio (VCR) of 5 and pH of 7.5.

Further purification was obtained by application of discontinuous diafiltration (DF) step consisting of adjustment of solution volume to its initial volume with tap water and re-VCR 5 at either pH of 7.5 or 6. Overall, four isolates were obtained as A) H2O/pH 7.5/25 °C/UF:DF pH 7.5:6, B) H2O/pH 7.5/25 °C/UF:DF pH 7.5:7.5, C) KCl/pH 7.5/25 °C/UF:DF pH 7.5:6, and D) KCl/pH 7.5/25 °C/UF:DF pH 7.5:7.5. They were freeze-dried, placed in aluminium pouches, sealed, and stored at 4°C prior to further examination. Maltodextrin of DE 15 (N-Lite® D) was provided by National Starch Food Products Division (New Jersey).

6.2. Preparation of Emulsions

Preparation of emulsions was based on the method by Taherian et al. (2011b) with slight modification. Fish oil was thawed at refrigerator temperature (4°C) 24h prior to emulsion preparation.

Protein solutions were prepared by addition of 210g double distilled (DD) water in a commercial blender and addition of 7.5g pea protein and 0.1g of sodium azide while blending at low speed following by blending at high speed for 3 minutes. Following total hydration (overnight at 4°C), protein solutions were then centrifuged at 4000G for 20 minutes to remove any non-soluble protein. The supernatant was recovered and pH was adjusted at 7 with the aid of an acid (HCL 0.2M) or a base (KOH 2M).

For maltodexterin, 22.5 g were also hydrated overnight in 210g double distilled water prior to preparation of secondary emulsions, using a similar approach than for the protein.

Primary emulsion was prepared by dispersion of 50 g of fish oil into the hydrated pea protein solution.

The procedure consisted of the preparation of a coarse emulsion by blending fish oil and hydrated protein for 3 min followed by high pressure homogenization (Emulsiflex-C5, Avestin, ON, Canada) at 5000 psi for 3 passes.

The secondary emulsions were prepared by addition of hydrated maltodextrin into the primary emulsion using the similar procedure of blending and homogenization shown in Figure 8. Final emulsion contains 10% fish oil, 1.5% protein and 4.5% maltodextrin. Prepared emulsions were tested right after preparation and within 15 and 30 days storage at room temperature.

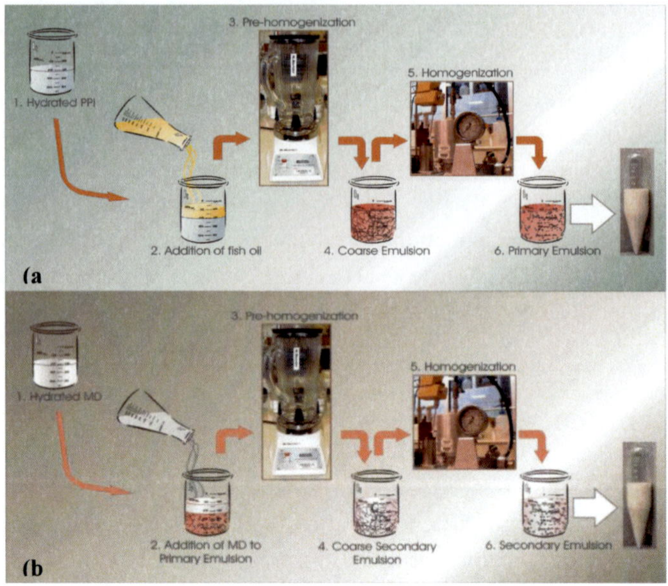

Figure 8. Schematic method for preparation of (a) primary and (b) secondary emulsion.

6.3. Flow and Dynamic Rheological Assessments

Measurement of rheological properties was based on the methods by Taherian et al. (2011b) using an AR1000 Rheometer (TA Instrument, New castle, DE, USA) with a 1.59 degree cone of 60mm diameter. Flow rheological properties of emulsions were measured at 0.1-100rad/s at 22°C. Shear viscosity (η^{γ}) as a function of shear rate, flow behaviour index (n) and consistency coefficient (m) at day-1, day-15 and day-30 were assessed using the power law model. Dynamic rheological properties tests were conducted at 22°C, 0.5 Pa stress and frequency of 1 to 25 rad/s to assess storage modulus (G'), loss modulus (G'') and delta degree (G''/G'). Each measurement was the result of 3 replicates and 3 tests for day-1, day-15 and day-30.

6.4. Determination of Droplet Size Distribution

The mean size diameter of emulsion droplets was determined using a dynamic light scattering instrument (Zetasizer Nano-ZS, Malvern instruments,

Worcs., UK). The measurements were conducted after preparation in duplicate and within 15 and 30 days.

6.5. Optical Microscopy

Prior to analysis, emulsion was gently agitated to ensure homogeneity and diluted 1: 10 using DD water.

A drop of diluted emulsion was then placed on a microscope slide and covered with a cover slip. The microstructure of emulsion was determined using a Nikon Eclipse E400 microscope (Nikon Corporation, Japan) and the Nikon ACT-1 version 2.12 software to process the images. Pictures were taken from eight different fields at magnification x1000 on each slide at day-1, 15 and 30.

6.6. Emulsion Stability

Emulsion stability was quantified based on the method provided by Taherian et al. (2007). Emulsions were subjected to stability test by pouring 6 ml of each emulsion into a flat-bottom cylindrical glass tube (100 mm height,16 mm internal diameter) and subjecting to an optical scanning instrument (Quick Scan, Coulter Crop., Miami, FL). The transmission of monochromatic light (λ 850 nm) from the emulsions was measured as a function of their height. Separation rate was quantified by conducting a total of 10 scans (each scan was repeated 5 times throughout 10 min) on each tube right after preparation of the emulsion, as well as after 1, 15 and 30 days of storage.

This quantification was based on the migration rate of the oil droplets from the bottom to the top of the sample which induces a progressive fall in concentration at the sample bottom (clarification) and therefore increases the transmission.

The resulting positive peaks were then transferred to Microsoft Excel and separation rate was calculated as slope of transmission mean values over 30 days storage (For complete information please refer to Mengual,Meunier, Cayré, Puech, and Snabre, 1999).

7. RESULTS AND DISCUSSION

7.1. Flow and Dynamic Rheological Properties of Prepared Emulsions

7.1.1. Flow Properties

Flow behaviours of emulsions were characterized by measuring the shear-rate dependent viscosity right after preparations, at day-15 and day-30. The shear rate dependence of apparent viscosities for commercial and membrane processed pea protein isolates are shown in Figure 9. The determination coefficients (R^2) for all measurements were more than 0.98 (not shown), indicating a high level of relation between measuring points.

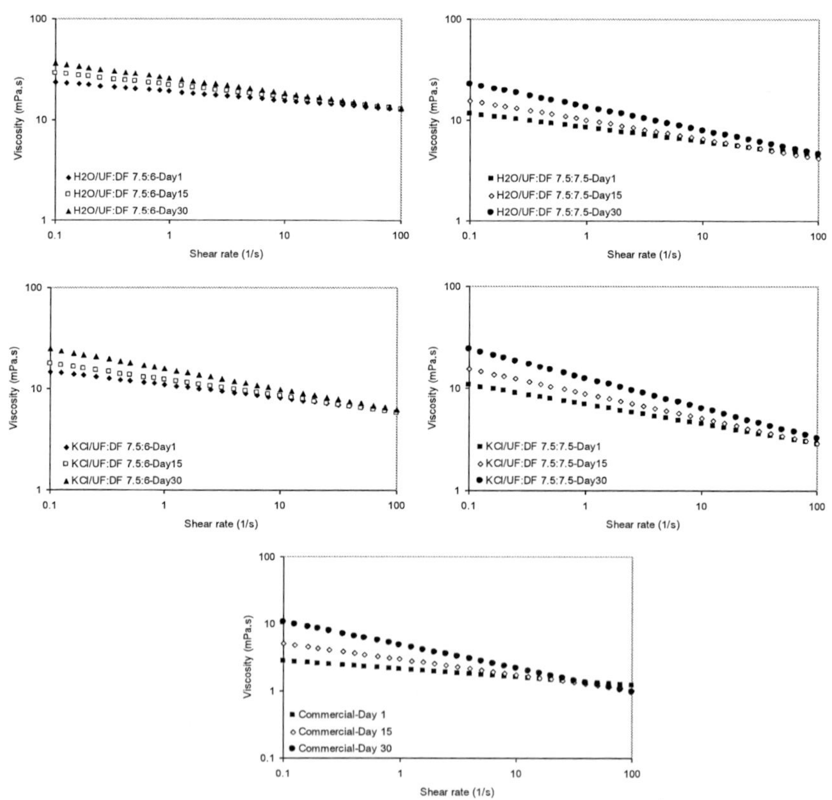

Figure 9. Extend of changes in apparent viscosity for prepared emulsions.

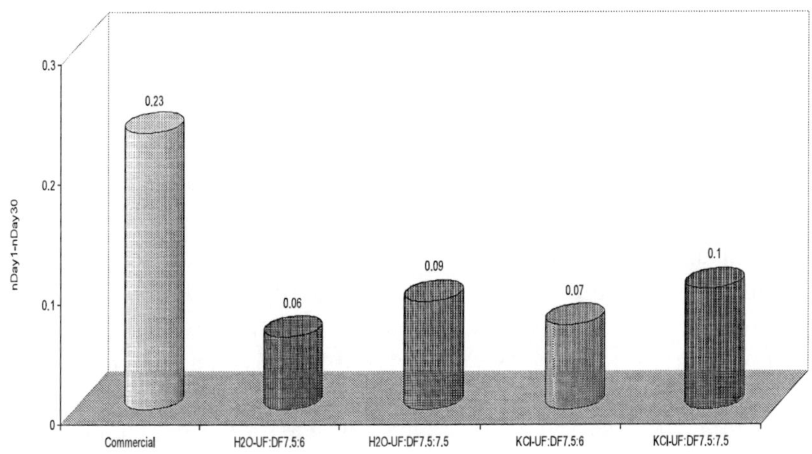

Figure 10. The extend of changes for flow behaviour index of tested emulsions.

All the emulsions showed shear thinning behaviour and the viscosity decreased along with increasing the shear rate. Rheological flow properties of all emulsions changed as influence of aging and showed an increase in apparent viscosities and a decrease in flow behaviour indexes. The degree of changes was different for different emulsions and the greatest changes were associated with the emulsion stabilized by commercial isolate-maltodextrin. Emulsions stabilized by H2O extracted at pH 7.5 and 25 °C, ultrafiltered at pH 7.5 and diafiltered at pH 6 (H2O/pH 7.5/25 °C/UF:DF pH 7.5:6) indicated the lowest changes for both apparent viscosity and flow behaviour index (Figure 10).

Previously reported data have shown that both increase in apparent viscosity and decrease in flow behaviour index of emulsion as an influence of aging is directly related to the onset of flocculation (Taherian et al., 2007; Vingerhoeds et al., 2009). When the flocs are small, increase in viscosity is minimal and flocculation is reversible upon application of slight shear. In the case of strong flocculation and coalescence of droplets the aggregation is irreversible and seems to be driven upon electrostatics interactions as was pointed out by Sillettia et al. (2007).

For commercial pea protein isolate-maltodextrin stabilized emulsions, the value of "n" decreased from 0.88 at day-1 to 0.65 at day-30 indicating unset of flocculation and coalescence for this emulsion. Emulsion stabilized by membrane processed isolate-maltodextrin (H2O/pH 7.5/25 °C/UF:DF pH 7.5:6), on the other hand, indicated the lowest decrease of "n" from 0.91 at day-1 to 0.85 at day-30 and showed, practically, Newtonian flow behavior for tested emulsion. This indicates that flocculation and coalescence was minimal for this emulsion. The decreases in "n" values for emulsions stabilized by H2O/pH 7.5/25 °C/UF:DF pH 7.5:7.5-maltodextrin, KCl/pH 7.5/25 °C/UF:DF pH 7.5:6-maltodextrin, and KCl/pH 7.5/25 °C/UF:DF pH 7.5:7.5-maltodextrin were from 0.86 to 0.77, from 0.87 to 0.80, and from 0.81 to 0.71, respectively, which were, considerably, lower than that of commercial pea protein isolate-maltodextrin stabilized emulsion.

Furthermore, changes in apparent viscosities of tested emulsions followed the same trends as for "n" values. Apparent viscosities (0.1/s) of emulsions prepared with H2O/pH 7.5/25 °C/UF:DF pH 7.5:6-maltodextrin, KCl/pH 7.5/25 °C/UF:DF pH 7.5:6-maltodextrin, H2O/pH 7.5/25 °C/UF:DF pH 7.5:7.5-maltodextrin, KCl/pH 7.5/25 °C/UF:DF pH 7.5:7.5-maltodextrin, and commercial isolate-maltodextrin were 23.46 mPa.s, 14.80 mPa.s, 11.52 mPa.s, 10.73 mPa.s and 2.76 mPa.s, respectively. Taherian et al. (2011a) found that the degree of decreases in apparent viscosity or flow behaviour index (n) depended upon, purity and solubility of the protein and since the isolates diafiltered at pH 6, contained lower amount of phytic acid, the rheological flow properties of these emulsions prepared with these isolates demonstrated lower changes.

7.1.2. Dynamic Properties

The dynamic rheological parameters (G', G'' and delta degree) of emulsions were evaluated. To insure storage modulus (G') and loss modulus (G") are reliable and accurate the rheological parameters were first measured as a function of stress amplitude at a fixed frequency. After establishment of linear viscoelastic region, measurements were then made at fixed stress amplitude (0.5 Pa) as a function of frequency. This way the linear region, where the dynamic parameters (G', G" and phase shift angle δ) are independent of the magnitude of applied stress, was investigated and proper measuring parameters were selected. The use of phase shift angle or delta degree (δ) in viscoelastic systems is based on the measurements of G' and G" modulus. Thus in a purely viscous system (i.e., water) δ is 90 °, and subsequently $G' = 0$ and $G'' = G^*$, where G^* is the complex modulus. Eventually, if the system is purely elastic δ is 0 °, and subsequently $G' = G^*$ and $G'' = 0$.

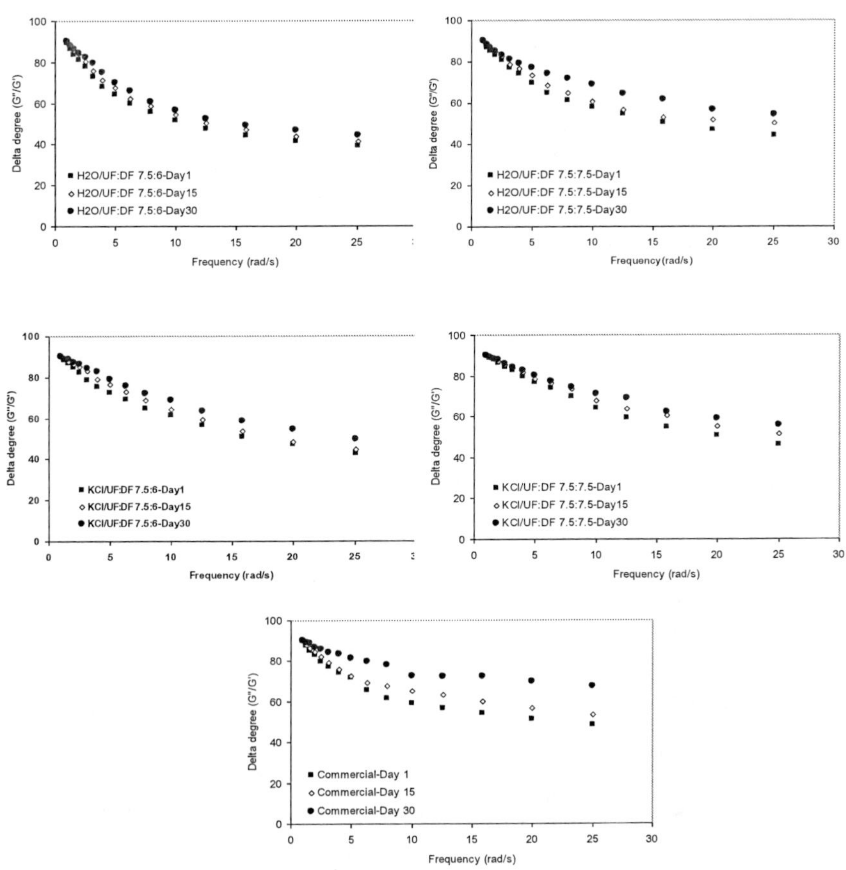

Figure 11. Extend of changes in delta degree for prepared emulsions.

Figure 11 compares the frequency dependence of delta degree (G''/G') for prepared emulsions at days 1, 15 and 30 and Figure 12 shows the extend of changes for delta degree of emulsions from day 1 to day 30 . The results indicate that the δ degree follow the same trend as shear viscosity as the delta degrees increase at all applied frequency level at different extends and showing similar trends for the extend of changes. The lowest values of δ degree at elevated level of applied frequency were associated with emulsions made with pea protein extracted in water and diafiltered at pH 6 and maltodextrin (39.06±1.01, day 1) followed by emulsions made with pea protein extracted in KCl solution and diafiltered at pH6 and maltodextrin

(42.44±1.05, day 1). The values of δ at elevated level of applied frequency for emulsions made with pea protein isolates extracted in water or KCl solution followed by diafiltration at pH 7.5 and maltodextrin as well as emulsions made with commercial isolate and maltodextrin at day 1 were 43.98±0.94, 46.15±1.07, and 48.2±1.72, respectively. The δ degrees for emulsions prepared with commercial isolate-maltodextrin and membrane processed isolates-maltodextrin were higher and slightly higher or lower than 45°, respectively, suggesting that liquid like behavior dominates emulsions made with commercial isolate-maltodextrin over solid like viscous behavior and vise versa. Therefore, the membrane processed isolates could provide greater emulsion stability when used as emulsifier.

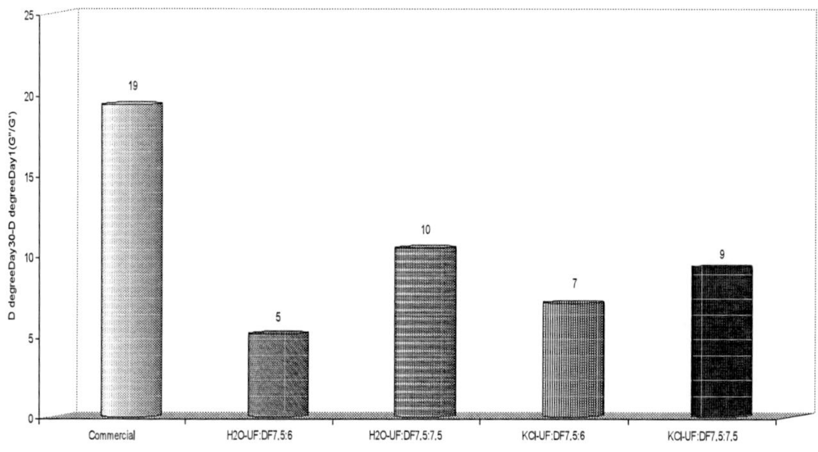

Figure 12. The extend of changes for delta degrees of tested emulsions.

7.2. Particle Size Growth, Degree of Coalescence and Emulsion Stability

In order to verify the changes in particle size, the mean average droplet size (Z-average size) was considered to compare emulsions stability and calculate the rate of coalescence (D_c). Emulsions were stored at room temperature and z-averages were measured at day-1, day-15, and day-30 in duplicate and a total of 3 tests for each measurement. Studies by Sherman (1983), Ye et al. (2004) Paraskevopoulou et al. (2005) and Taherian et al. (2008), Taherian et al. (2011b) showed that the rate of coalescence of emulsion droplets (D_c) mainly follows the first-order kinetics (Eq. 2).

$$N_t = N_0 \exp\left(-D_c t\right) \tag{2}$$

where N_0 and N_t are the numbers of droplets per unit volume of emulsion initially and time t, respectively, and D_c is the rate of droplets coalescence. Using mean average droplet size measured after preparation, day-15 and 30 the rate of coalescence (D_c) can be determined by plotting $3(\ln (D_t/D_0))$ versus time (t) using Equation (3):

$$\ln D_t = \ln D_0 + \frac{D_c t}{3} \tag{3}$$

where D_0 and D_t are the average droplet sizes initially and at time t, respectively.

As are shown in Figure 13, emulsion stabilized by commercial pea protein isolate-maltodextrin and emulsion made with isolate extracted in water and diafiltered at pH 6 indicated the highest and lowest rates of particle growth and coalescences, respectively. Other emulsions made with membrane processed pea protein isolates also show lower degree of particle growth and coalescences than for the emulsion made from the commercial protein. The fact that the apparent viscosity (0.1/s) of emulsions made with water extracted and diafiltered at pH6 isolate had the highest viscosity compare to the other emulsions suggests that steric interactions have played an essential role in retarding the droplets coalescence for this emulsion. This is also true for all other membrane processed stabilized emulsions compare to commercial isolate stabilized emulsion.

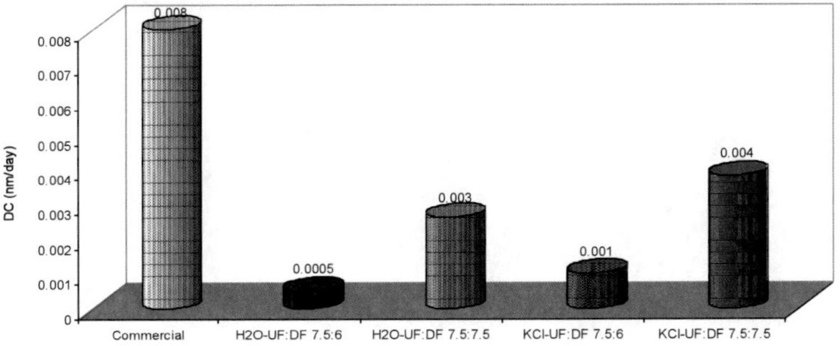

Figure 13. Degree of coalescence for prepared emulsions after 30 days of aging.

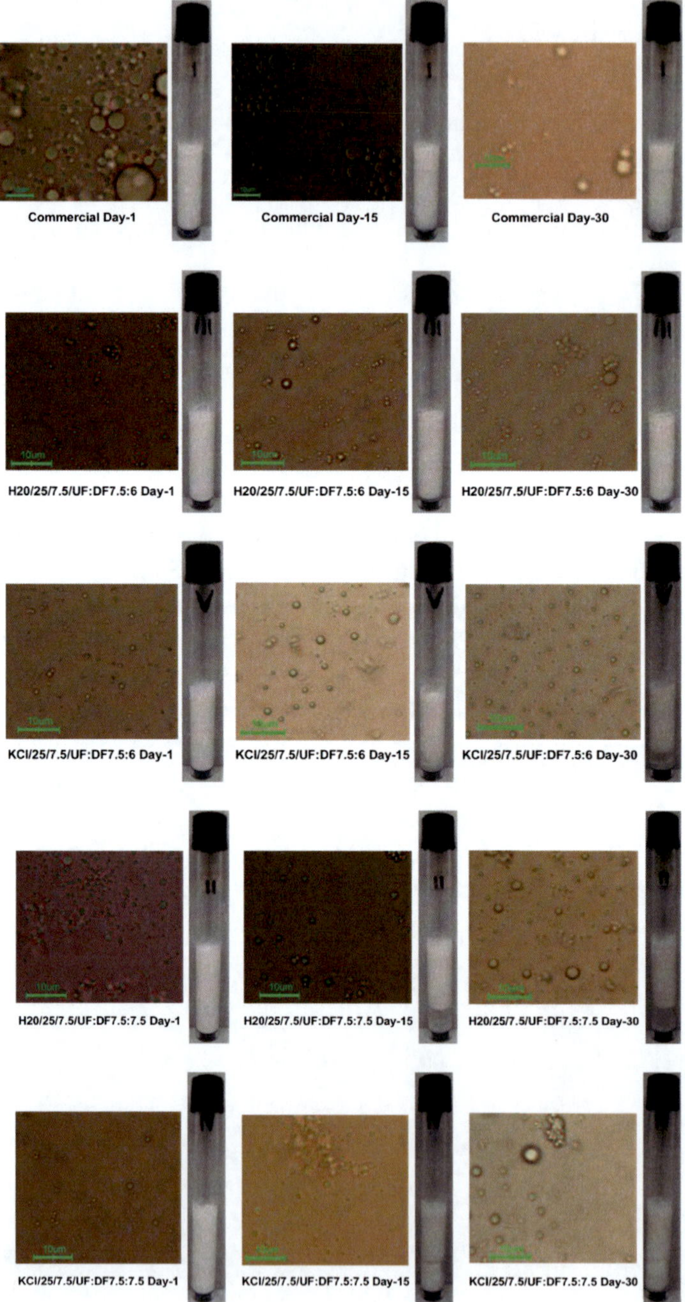

Figure 14. Optical microscopy and physical separation of tested emulsions at days 1, 15, and 30.

The microstructures of studied emulsions along with visual observations in screw cap test tubes obtained after preparation, day-15 and day-30 are compared in Figure 14. As can be observed, the micrographs display evidence of aggregated droplets of commercial pea protein isolate coated droplets. The micrographs for emulsions prepared with conjugation of membrane processed isolates-maltodextrin show lower degree of coalescence and confirm the segregation effect of the conjugated protein-polysaccharide interfacial film.

The lesser degree of coalescence for membrane processed isolates stabilized emulsions could also be related to both changes in the conformation of the protein molecule and the net charge of the adsorbed protein layers at the interface. As Pearce and Kinsella (1987) stated, the ability of the protein to be adsorbed at the surface of oil droplets depends upon its capacity to unfold and spread over the interface to stabilize the new area created. As a result, the conformational changes of the protein molecule are extremely important, because they are related to properties such as surface hydrophobicity, protein flexibility, solubility, degree of disulphide bonds, degree of hydrogen interactions and other stabilizing forces. This also suggest that the thickness of the continuous phase around the droplets (interstitial continuous phase) was enough to minimize contact between droplet films, and interfacial fluctuations did not lead to the exclusion of interstitial waterMc Clements, 1999 D.J. Mc Clements, Food emulsions. Principles, practice and techniques, CRC Press, New York (1999).

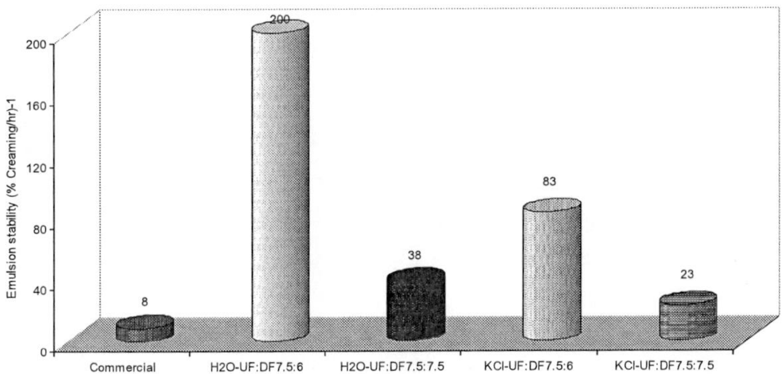

Figure 15. Comparison of stability of emulsions prepared with commercial and membrane processed pea protein isolates.

The stability of emulsions is compared in Figure 15. As the highest rate of aggregation is associated to commercial pea protein isolate stabilized emulsion, the results are in excellent agreement with previously mentioned coalescence rate presented in Figure 13. Therefore, the higher stabilities of emulsions prepared with membrane processed isolates could be related to the higher solubility and their higher capacity to unfold and cover larger parts of the surfaces.

GENERAL CONCLUSIONS

With the worldwide total production of 10.3 million metric tonnes of pea (Simsek et al., 2009) there is a strong need for exploiting the potential of this product as food ingredients to supply efficient protein and open a new door to industrial application of this protein. Despite of the fact that peas are excellent source of proteins, carbohydrates and essential minerals and that pea proteins show a well balanced profile of amino acids, especially a high content in lysine (Nunes et al., 2006), its utility to humans is limited by the presence of antinutrients, such as enzyme inhibitors, lectins, phenolic compounds, phytates, oligosaccharides, and saponins. Several processes could impact the reduction of these antinutrients and among them membrane processing in combination with certain pre-treatments has great potential. Consequently, there are needs to conduct studies with a particular concern on using pea protein isolates as a source of nutritive and functional ingredient for development of novel food products.

ACKNOWLEDGMENTS

The authors would like to thanks Ms. Marion Faure, Ms. Hélène Drolet and Mr. Denis Ippersiel for their technical assistances, as well as, Agriculture and Agrifood Canada for financial supports.

REFERENCES

Abd El-Hady, E.A., and Habiba, R.A. (2003). Effect of soaking and extrusion conditions on antinutrients and protein digestibility of legume seeds, Lebensm.-*Wiss. U.-Technol.*, 36, p. 285-293.

Adebiyi A.P., and Aluko R. E., 2011. Functional properties of protein fractions obtained from commercial yellow field pea (Pisum sativum L.) seed protein isolate. *Food Chemistry*, 128, 902–908.

Agboola S.O., Mofolasayo O.A., Watts B.M., Aluko R.E. (2010). Functional properties of yellow field pea (Pisum sativum L.) seed flours and the in vitro bioactive properties of their polyphenols. *Food Research International* 43, 582–588.

Agriculture and Agri-Food Canada, Pulses and Special Crops, 2007. http://www4.agr.gc.ca/AAFC-AAC/display-afficher.do?id=1174597774743andlang=eng.

Ali, F., Ippersiel, D., Lamarche, F., and Mondor, M. (2010), Characterization of low-phytate soy protein isolates produced by membrane technologies, *Innovative Food Science and Emerging Technologies*, 11(1), p. 162-168.

Alonso, R., Orue, E., and Marzo, F. (1998), Effects of extrusion and conventional processing methods on protein and antinutritional factor contents in pea seeds, *Food Chemistry*, 63(4), p. 505-512.

Aluko, R. E., Mofolasayo, O. A., and Watts, B. M. (2009). Emulsifying and foaming properties of commercial yellow pea (Pisum sativum L.) seed flours. *Journal of Agricultural and Food Chemistry*, 57, 9793–9800.

Bautista-Teruel, M.N., Eusebio, P.S., Welsh, T.P., 2003. Utilization of feed pea, *Pisum sativum*, meal as a protein source in practical diets for juvenile tiger shrimp, *Penaeus monodon. Aquaculture* 225, 121–131.

Beal, L., and Mehta, T. (1985). Zinc and phytate distribution in peas-Influence of heat treatment, germination, pH, substrate, and phosphorus on pea phytate and phytase, *Journal of Food Science*, 50, p. 96-115.

Belfort, G., Davis, R.H., and Zydney, A.L. (1994). The behaviour of suspensions and macromolecular solutions in crossflow microfiltration. *Journal of Membrane Science*, 96, 1-58.

Bishnoi, S., and Khetarpaul, N. (1994). Saponin content and trypsin inhibitor of pea cultivars: Effect of domestic processing and cooking methods, *Journal of Food Science and Technology*, 31(1), p. 73-76.

Bishnoi, S., Khetarpaul, N., and Yadav, R.K. (1994). Effect of domestic processing and cooking methods on phytic acid and polyphenol contents of pea cultivars (Pisum sativum), *Plant Foods for Human Nutrition*, 45, p. 381-388.

Bora, P. S., Clark, J. B., and Powers, J. R. (1994). Heat-induced gelation of pea (Pisum sativum) mixed globulins, vicilin and legumin. *Journal of Food Science*, 59, 594–596.

Bouaouina H., Desrumaux A., Loisel C. and Legrand J., 2006. Functional properties of whey proteins as affected by dynamic high-pressure treatment. *International Dairy Journal* 16, 275-284.

Brooks, J.R., and Morr, C.V. (1985). Effect of phytate removal treatments upon the molecular weight and subunit composition of major soy protein fractions. *Journal of Agricultural and Food Chemistry*, 33, p. 1128-1132.

Campos-Vega R., Loarca-Pina G., Oomah B.D. 2010. Review, Minor components of pulses and their potential impact on human health. *Food Research International* 43, 461–482.

Champ, M. M. (2002). Non-nutrient bioactive substances of pulses. British *Journal of Nutrition*, 88(Suppl. 3), S307–S319.

Chavan U.D., McKenzie D.B., Shahidi F., 2001. Protein classification of beach pea (Lathyrus maritimus L.). *Food Chemistry*, 75, 145–153.

Cheryan, M. (1998). *Ultrafiltration and microfiltration handbook.* Chicago: Technomic Publ.

Chung, K.T., Wong, T.Y., Wei, C. I., Huang, Y. W., and Lin, Y. (1998). Tannins and human health: A review. Critical Reviews in *Food Science and Nutrition*, 38, 421–464.

Cruz-Suarez, L.E., Marie, D.M., Salazar, M.T., McCallum, I.M., Hickling, D., 2001. Assessment of differently processed feed pea *Pisum sativum* meals and canola meal *Brassica* sp. in diets for blue shrimp *Litopenaeusstylirostris*. *Aquaculture* 196, 87–104.

Cuperus, F.P., and Nijhuis, H.H., 1993. Applications of membrane technology to food processing. *Trends in Food Science and Technology*, 7, p. 277–282.

Davis, D.A., Arnold, C.R., Mc Callum, I., 2002. Nutritional value of feed peas (*Pisum sativum*) in practical diet formulations for *Litopenaeus vannamei*. *Aquacult. Nutr.* 8, 87–94.

de Morais Coutinho C., Chiu M.C., Basso R.C., Badan Ribeiro A.P., Guaraldo Gonçalves L.A., and Viotto L.A. 2009. State of art of the application of membrane technology to vegetable oils: A review, Food *Research International*, 42, p. 536-550.

Derbyshire, E., Wright, D. J., and Boulter, D., 1976. Legumin and vicilin, storage proteins of legume seeds. *Phytochemistry*, 15, 3–24.

Dervas G., Doxastakis G., Hadjisavva-Zinoviadi S., Triantafillakos N., 1999. Lupin flour addition to wheat flour doughs and effect on rheological properties. *Food Chemistry*, 66, 67–73.

Doxastakis, G., 2000. *Lupin seed proteins*. In G. Doxastakis, and V. Kiosseoglou (Eds.), Novel macromolecules in food systems (pp. 7–38). Amsterdam: Elsevier, 7–38.

El-Adawy T.A., Rahma E.H., El-Bedawey A.A., and El-Beltagy A.E., 2003. Nutritional potential and functional properties of germinated mung bean, pea and lentil seeds, *Plant Foods for Human Nutrition*, 58, p. 1-13.

Eusebio P.S., 1991. Effect of dehulling on the nutritive value of some leguminous seeds as protein sources for tiger prawn *Penaeus monodon* juveniles. *Aquaculture* 99, 297–308.

FAO, 1970. Amino acid content of foods and biological data on proteins. *Nutr. Studies Rep.* No. 24. FAO. Rome.

Fredrikson M., Biot P., Larsson Alminger M., Carlsson N.-G., and Sandberg A.-S., 2001. Production Process for High-Quality Pea-Protein Isolate with Low Content of Oligosaccharides and Payette, *Journal of Agricultural and Food Chemistry*, 49, p. 1208-1212.

Fuhrmeister H., Meuser F., 2003. Impact of processing on functional properties of protein products from wrinkled peas. *Journal of Food Engineering*, 56, p. 119-129.

Gao Y., Shang C., Saghai Maroof M.A., Biyashev R.M., Grabau E.A., Kwanyuen P., Burton J.W., Buss G.R., 2007. A modified colorimetric method for phytic acid analysis in soybean, *Crop Science*, 47, p. 1797-1803.

Gatel F. and Grosjean F., 1990. Composition and nutritive value of peas for pigs: a review of European results. *Livestock Production Science,* 26, 155-175 155.

Gueguen, J., 1983. in Plant proteins for human food (Bodwell C.E. and Petiti, L. Eds.), Martinus Nijhoff/Dr. W. Junk Publisher, *Hague*, 63-68.

Habiba R.A. (2002). Changes in anti-nutrients, protein solubility, digestibility, and HCl-extractability of ash and phosphorus in vegetable peas as affected by cooking methods. *Food Chemistry* 77, 187–192.

Igbasan F.A., Guenter W., 1996. The evaluation and enhancement of the nutritive value of yellow-, green- and brown-seeded pea cultivars for unpelleted diets given to broiler Chickens. *Animal Feed Science Technology*, 63, 9-24.

Jaffé, W.G., Moreno, R., and Wallis, V. (1973), Amylase inhibitors in legume seeds, *Nutrition Reports International*, 7, p. 169-174.

Jianmei Y., Ahmedna M. and Goktepe I., 2007. Peanut protein concentrate: Production and functional properties as affected by processing. *Food Chemistry* 103, 121-129.

Jourdan, G. A., Norena, C. P. Z., and Brandelli, A. (2007), Inactivation of trypsin inhibitor activity from Brazilian varieties of beans (Phaseolus vulgaris L.)., *Food Science and Technology International*, 13(3), p. 195-198.

Khattab R.Y., Arntfield S.D., 2009. Nutritional quality of legume seeds as affected by some physical treatments 2. Antinutritional factors. LWT - *Food Science and Technology*, 42, 1113–1118.

Kinsella, J. E. (1979). Functional properties of soy proteins. *Journal of the American Oil Chemists' Society*, 56, 242–258.

Kramadhati, N.N., Mondor, M., and Moresoli, C. (2002), Evaluation of the shear-induced diffusion model for the microfiltration of polydisperse feed suspension, *Separation and Purification Technology*, 27, 1, p. 11-24.

Kumaraguru Vasagam K.P., Balasubramanian T., Venkatesan R., 2007. Apparent digestibility of differently processed grain legumes, cow pea and mung bean in black tiger shrimp, *Penaeus monodon* Fabricius and associated histological anomalies in hepatopancreas and midgut. *Animal Feed Science and Technology*, 132, 250–266.

Lallés J.P., 1993. Nutritional and antinutritional aspects of soyabean and field pea proteins used in veal calf production: a review, *Livestock Production Science*, 34, 181-202.

Lin, L., Rhee, K. C., and Koseoglu, S. S. (1997). Bench-scale membrane degumming of crude vegetable oil: Process optimization, *Journal of Membrane Science*, 134, p. 101-108.

Liu S., Elmer C., Low N.H., Nickerson M.T. (2010). Effect of pH on the functional behaviour of pea protein isolate–gum Arabic complexes. *Food Research International* 43, 489–495.

Liu Z.-H., Cheng F.-M., Cheng W.-D., Zhang G.-P., 2005. Positional variations in phytic acid and protein content within a panicle of japonica rice. *Journal of Cereal Science* 41, 297–303.

Ma, Z., Boye, J.I., Simpson, B.K., Prasher, S.O., Monpetit, D., and Malcolmson, L. (2011), Thermal processing effects on the functional properties and microstructure of lentil, *Food Research International*, 44, p. 2534-2544.

Makri E., Papalamprou E., Doxastakis G. (2005). Study of functional properties of seed storage proteins from indigenous European legume crops (lupin, pea, broad bean) in admixture with polysaccharides*Food Hydrocolloids* 19, 583–594.

Makri E., Papalamprou E., Doxastakis G. (2005). Study of functional properties of seed storage proteins from indigenous European legume crops (lupin, pea, broad bean) in admixture with polysaccharides *Food Hydrocolloids* 19, 583–594.

Marshall, A. D., and Daufin, G. (1995). Physico-chemical aspects of membrane fouling by dairy fluids. In IDF Special Issue 9504 (Ed.), *Fouling and cleaning in Federation.*

Mengual,O., Meunier,G., Cayré, I., Puech,K.,andSnabre,P. (1999). Characterisation of instability of concentrated dispersions by a new optical analyser: the TURBISCAN MA 1000. Colloids and SurfacesA; *Physicochemical and Engineering Aspects*, 152,111-123

Meuser, F., Pahne, N., and Möller, M. (1997). Yield of starch and byproducts in the processing of different varieties of wrinkled peas on a pilot scale, *Cereal Chemistry*, 74, p. 364-370.

Mondor, M., Aksay, S., Drolet, H., Roufik, S., Farnworth, E., and Boye, J.I. (2009b), Influence of processing on composition and antinutritional factors of chickpea protein concentrates produced by isoelectric precipitation and ultrafiltration, *Innovative Food Science and Emerging Technologies*, 10, p. 342-347.

Mondor, M., Ippersiel, D., Lamarche, F., and Boye, J.I. (2004), Production of soy protein concentrates using a combination of electro-acidification and ultrafiltration, *Journal of Agricultural and Food Chemistry*, 52(23), p. 6991-6996.

Mondor, M., Tuyishime, O., and Drolet, H. (2009a), Influence of hollow-fibre lumen diameter on the purification of pea extracts by ultrafiltration, Proceeding of the: "*Journée d'information   scientifique et technique en génie agroalimentaire*", St-Hyacinthe, Québec, (Canada), 8 pages.

Mubarak, A.E., 2005. Nutritional composition and antinutritional factors of mung bean seeds (*Phaseolus aureus*) as affected by some home traditional processes. *Food Chem.* 89, 489–495.

Mwasaru M.A., Muhammad K., Bakar J., Che Man Y.B. (2000). Infuence of altered solvent environment on the functionality of pigeonpea (Cajanus cajan) and cowpea (Vigna unguiculata) protein isolates. *Food Chemistry*, 71, 157-165.

Norton, G. (1991). Proteinase inhibitors. In J. P. F. D' Mello, C. M. Duffus, and J. H. Duffus (Eds.), Toxic substances in crop plants (pp. 68–106). Cambridge: *The Royal Society of Chemistry.*

Nunes M.C., Raymundo A., Sousac I. (2006). Rheological behaviour and microstructure of pea protein/k-carrageenan/starch gels with different setting conditions *Food Hydrocolloids*, 20, 106–113.

O'Kane, F. E., Vereijken, J. M., Gruppen, H., and van Boekel, M. A. J. S. (2005). Gelation behaviour of protein isolates extracted from 5 cultivars of Pisum sativum L. *Journal of Food Science*, 70, 132–137.

Omosaiye, O., and Cheryan, M. (1979). Low-phytate, full-fat soy protein product by ultrafiltration of aqueous extracts of whole soybeans, *Cereal Chemistry*, 56, p. 58-62.

Owusu-Ansah Y. J. and Mc Curdy S. M. (1991). Pea proteins: A review of chemistry, technology of production, and utilization. *Food Reviews International*, 7, 1, 103-134.

Owusu-Ansah, Y.J., and Mc Curdy, S.M. (1991), Pea proteins: A review of chemistry, technology of production, and utilization, *Food Reviews Internationals*, 7(1), p. 103-134.

Owusu-Apenten, R.K. (2002). Food protein analysis: *Quantitative effects on processing*. Marcel Dekker, New York.

Pearce K.N. and Kinsella J.E. (1987). Emulsifying properties of proteins: evaluation of a turbidimetric technique. *Journal of Agricultural and Food Chemistry* 26, 716–723.

Phillippy, B. Q. (2003). Inositol phosphates in foods. *Advances in Food and Nutrition Research*, 45, 1–60.

Rangel, A., Domont, G. B., Pedrosa, C., and Ferreira, S. T. (2003). Functional properties of purified vicilins from cowpea protein isolate. *Journal of Agricultural and Food Chemistry*, 51, 5792–5797.

Reddy, N. R., Pierson, M. D., Sathe, S. K., and Salunkhe, D. K. (1985). Dry bean tannins– a review of nutritional implications. *Journal of the American Oil Chemists Society*, 62, 541–549.

Rivas-Vega, M.E., Goytort´ua-Bores, E., Ezquerra-Brauer, J.M., Salazar-Garc´ıa, M.G., Cruz-Su´arez, L.E., Nolasco, H., Civera-Cerecedo, R., 2006. Nutritional value of cowpea (*Vigna unguiculata* L.Walp) meals as ingredients in diets for Pacific white shrimp (*Litopenaeus vanname* Boone*i*). *Food Chem.* 97, 41–49.

Roman, A. V., Bender, A. E., and Morton, I. D. (1987), Formulation and processing of a weaning food based on rice, cowpeas and skim milk powder, *Human Nutrition: Food Science and Nutrition*, 41F(1), p. 15-22.

Sandberg, A. S. (2002). Bioavailability of minerals in legumes. *British Journal of Nutrition*, 88, 281–285.

Scott, K. (2003). Handbook of industrial membranes. Oxford: Elsevier.

Shand P.J., Ya H., Pietrasik Z., Wanasundara P.K.J.P.D. (2007). Physicochemical and textural properties of heat-induced pea protein isolate gels. *Food Chemistry*, 102, 1119–1130.

Shekib, L.A., El-Iraqui, S.M., and Abo-Bakr, T.M. (1988), Studies on amylase inhibitors in some Egyptian legume seeds, *Plant Foods for Human Nutrition*, 38, 325-332.

Shi, J., Arunasalam, K., Yeung, D., Kakuda, Y., Mittal, G., and Jiang, Y. (2004), Saponins from edible legumes: Chemistry, processing, and health benefits, *Journal of Medicinal Food*, 7, 67–78.

Simsek S., Tulbek M.C., Yao Y., Schatz B. (2009). Starch characteristics of dry peas (Pisum sativum L.) grown in the USA. *Food Chemistry* 115, 832–838.

Stamatakis, K., and Tien, C. (1993), A simple model of cross-flow filtration based on particle adhesion. *AIChE Journal*, 39(8), 1292-1302.

Strathmann, H. (1990). *Synthetic membranes and their preparation*. In M. C. Porter (Ed.), Handbook of industrial membrane technology. New Jersey: Noyes Publications.

Sudaryono, A., Tsvetnenko, E., Hutabarat, J., Supriharyono, E., 1999. Lupin ingredients in shrimp (*Penaeus monodon*) diets: influence of lupin species and types of meals. *Aquaculture* 171, 121–133.

Sunday, G., Monday, A., Juliet, E., 2001. Evaluation of selected food attributes of four advanced lines of ungerminated and germinated Nigerian cowpea (*Vigna unguiculata* (L.) Walp). *Plant Foods Hum. Nutr.* 56, 61–73.

Taherian A.R. ☐, Mondor M., Labranche J., Drolet H., Ippersiel D., Lamarche F., 2011. Comparative study of functional properties of commercial and membrane processed yellow pea protein isolates. *Food Research International*, 44, 2505–2514.

Taherian, A.R., Mondor, M., Labranche, J., Drolet, H., Ippersiel, D., and Lamarche, F. (2011), Comparative study of functional properties of commercial and membrane processed yellow pea protein isolates, *Food Research International*, 44, p. 2505-2514.

Taherian,A.R.,Fustier,P., and Ramaswamy,H.S. (2007) Steady and dynamic shear rheological properties and stability of non- flocculated and flocculated beverage emulsions. International *Journal of Food Properties*, 10, 915–934.

Tsumura K., Saito T., Tsuge K., Ashid A. H., Kugimiya W. and Inouye K., 2005. Functional properties of soy protein hydrolysates obtained by selective proteolysis. *LWT Food Science and Technology* 38, 255-261.

Uwaegbute, A.C., Iroegbu, C.U., Eke, O., 2000. Chemical and sensory evaluation of germinated cow peas (*Vigna unguiculata*) and their products. *Food Chem.* 68, 141–146.

van der Poel, A.T.F.B., Stolp, W., and van Zuilichem, D.J. (1992), Twin-screw extrusion of two pea varieties: Effects of temperature and moisture level on antinutritional factors and protein dispersibility, *Journal of the Science of Food and Agriculture*, 58, p. 83-87.

Vingerhoeds M.H., Silletti E., de Groot J., Schipper R.G., van Aken G.A. (2009). Relating the effect of saliva-induced emulsion flocculation on rheological properties and retention on the tongue surface with sensory perception. *Food Hydrocolloids* 23, 773–785.

Vose, J.R. (1980), Production and Functionality of Starches and Protein Isolates from Legume Seeds (Field Peas and Horsebeans), *Cereal Chemistry*, 57(6), p. 406-410.

Voutsinas L.P., Cheung E., and Nakai S., 1983. Relationships of hydrophobicity to emulsifying properties of heat denatured proteins. *Journal of Food Science*, 48, 26-32.

Wang N., Hatcher D.W., Warkentin T.D., Toews R. (2010). Effect of cultivar and environment on physicochemical and cooking characteristics of field pea (Pisum sativum). *Food Chemistry*, 118, 109–115.

Wang, N., and Daun, J. K. (2005), Determination of cooking times of pulses using an automated Mattson cooker apparatus, *Journal of the Science of Food and Agriculture*, 85, p. 1631-1635.

Wang, N., Daun, J. K., and Malcolmson, L. J. (2003), Relationship between physicochemical and cooking properties, and effects of cooking on antinutrients, of yellow field peas (Pisum sativum), *Journal of the Science of Food and Agriculture*, 83(12), p. 1228-1237.

Wang, N., Hatcher, D.W., and Gawalko, E.J. (2008), Effect of variety and processing on nutrients and certain anti-nutrients in field peas (Pisum sativum), *Food Chemistry*, 111, p. 132-138.

Wang, T. L., Domoney, C., Hedley, C. L., Casey, R., and Grusak, M. A. (2003). Can we improve the nutritional quality of legume seeds? *Plant Physiology*, 131, 886–891.

Yufei H., Youru H., Aiyong Q. and Xiaoya L., 2005. Properties of soy protein isolate prepared from aqueous alcohol washed soy flakes. *Food Research International* 38, 273-279.

Zeman, L.J., and Zydney, A.L. (1996). *Microfiltration and ultrafiltration: Principles and applications*. New York: Marcel Dekker, Inc.

In: Peas
Editors: A. Comstock and B. Lothrop

ISBN: 978-1-61942-866-9
© 2012 Nova Science Publishers, Inc.

Chapter 2

MARKET CLASS, CULTIVAR, LOCATION, AND CROP YEAR EFFECTS ON THE VOLATILE FLAVOUR COMPOSITION OF FIELD PEA CULTIVARS

Sorayya Azarnia[1], Joyce I. Boye[1], Tom Warkentin[2] and Linda Malcolmson[3]

[1]Agriculture and Agri-Food Canada, Food Research and Development Centre, St-Hyacinthe, QC, Canada,
[2]Crop Development Centre, University of Saskatchewan, Agriculture Building, Saskatoon, Saskatchewan, Canada
[3]Canadian International Grains Institute, Winnipeg, Manitoba, Canada

ABSTRACT

In this study, the volatile flavour profile of 24 pea cultivars were identified using head space solid phase microextraction gas chromatography coupled with mass spectrometry. Cultivars from several breeding programs were utilized with the purpose of understanding the genetic diversity of the cultivars in relation to differences in their flavour compounds. Furthermore, the effect of market class, cultivar, location, and crop year on the flavour characteristic of peas was evaluated. Cultivars were selected from 4 market classes of peas (i.e., yellow, green, dun, and marrowfat) grown in four locations (i.e., Sutherland (SUT); Meath Park (MPK); Pasqua (PAS); Wilkie (WIL) in Saskatoon, Canada in different crop years. Significant differences ($P < 0.01$) in volatile flavours were observed between field pea cultivars and the concentration of flavour compounds was significantly affected by the parameters studied. The 2008 crops had higher TVC, alcohols, ketones, pyrazines, and hydrocarbons than those from the year of 2009. In contrast, the 2009 crops had higher concentrations of aldehydes, sulfur compounds, esters, and terpenes. Overall, Cooper and Rambo had the highest mean value of TVC, whereas Kaspa had the lowest mean value of TVC. Peas grown in WIL and MPK had higher TVC compared to those from other locations. The highest mean value of TVC was observed in peas from the marrowfat-market class, whereas peas from the dun-market class had the lowest TVC. The lowest value of alcohols, aldehydes, ketones, sulphur compounds, hydrocarbons and pyrazines was found in the marrowfat-market class. In contrast, the dun-market class had the highest value of these compounds except for ketones and alcohols. Ketones were found in the highest amount in yellow- and green-market classes, and alcohols were observed at the highest level in green-market class. 2-Ethyl-1-hexanol, hexanal, 2-butanone, dimethyl sulfide, styrene, 2,3-diethyl-5-methyl pyrazine, ethyl acetate and hexanoic acid,methyl ester were the most abundant flavour compounds found in peas grown in both crop years. As the taste and flavour of peas could be affected by varying concentrations of different volatile compounds, these findings could be useful in the identification of specific cultivars for different food applications.

INTRODUCTION

Although pulses such as peas are known to be healthy legumes with nutritional benefits (Sathe 2002; Tharanathan and Mahadevamma 2003; de Almeida Costa et al. 2006), their consumption in some parts of the world, especially in Western countries is low. In addition to the length of time it takes to cook them, undesirable flavours such as green, grassy and beany flavours could be obstacles to consumer acceptance.

As different flavour compounds have different characteristics, variations in the flavour profile of peas could affect their taste and flavour. Furthermore, knowledge of the impact of parameters such as cultivar, location, crop year on the flavour properties of peas could help in identifying ideal cultivars to utilize in different food applications. Information about the impact of the above parameters on the flavour profile of peas is, however, lacking in the literature.

Canada is the world's largest producer and exporter of field peas (AAFC 2006) and the province of Saskatchewan produces 78% of the peas grown in the country. This study, therefore, was carried out to evaluate the impact of market class, cultivar, location and crop year on the flavour profile of field peas grown in the province of Saskatchewan in Canada.

Cultivars were selected from 4 market classes of field pea, i.e. yellow, green, dun, and marrowfat. Within a given market class, cultivars were generally selected based on the largest quantities produced, as these are the ones most likely to reach human consumption markets around the world.

For this study, the largest number of cultivars was from the yellow cotyledon market class, as this class encompasses about two-thirds of the field pea production in western Canada. Yellow cotyledon peas enter human consumption markets in many countries, particularly in the Indian sub-continent where they are used as dhal and flour. The next largest number of cultivars was from the green cotyledon market class which makes up most of the remaining production. Green cotyledon peas enter human consumption markets in many countries, particularly in South America where they are used as either whole, split, or canned peas. Marrowfat field pea production is quite small in western Canada with about 20,000 tonnes produced annually. The UK is the other significant producer (personal communication, Walker Seeds, Inc.). Dun field pea production is the prominent market class in Australia. Dun peas from Australia are sold into human consumption markets in India where they are typically dehulled and used as dhal.

MATERIALS AND METHODS

Materials

Chemicals and pure volatile standards were purchased from Sigma–Aldrich (Oakville, ON, Canada). Carboxen–polydimethylsiloxane (CAR /PDMS, 85 μm, Supelco, Oakville, ON, Canada) was selected as the SPME fibre for the GC analysis.

Yellow- (Crop Development Center (CDC) Golden, Eclipse, Cutlass, Canstar, CDC Bronco, CDC Centennial, CDC Meadow, CDC Mozart, CDC Tucker, DS Admiral, Reward, SW Midas), green- (CDC Striker, Cooper, Nitouche, CDC Patrick (1434-20), SW Parade, SW Sergeant), marrowfat- (MFR042, MFR043, MFS041, Rambo) and dun- (CDC Dundurn, Kaspa) market class were evaluated in this study. The field pea cultivars were grown in two replicates under uniform conditions using recommended agronomic practices for field pea on land managed by the Crop Development

Centre/University of Saskatchewan, Canada. The cultivars were grown at 4 different locations (i.e. Sutherland, SUT; Meath Park, MPK; Pasqua, PAS; Wilkie, WIL, near Saskatoon, Saskatchewan, Canada) in crop years of 2008 and 2009.

Methods

Standard Preparation

Pure volatile standards used in this study were selected on the basis of our earlier study (Azarnia et al. 2011). 2-Pentanone-d_5 was used as a daily reference standard during each GC run.

Preparation Of Milled-Whole Seeds

Pea flours were prepared using a small coffee mill, followed by sieving with a 106 µm sieve. The sieved material was used for analysis.

Solid Phase Microextraction Gas Chromatography Mass Spectrometry (HS-SPME-GC/MS) Analyses

Volatile compounds in the pea flours were extracted and identified using the HS-SPME-GC/MS procedure described in Azarnia et al. (2011). Samples (3 g) were extracted at 50 °C for 30 min using HS-SPME. Determination of volatile compounds was carried out with a Varian CP-3800 gas chromatograph (Palo Alto, CA). Desorption of volatile compounds were performed at 300 °C for 3 min using a split/splitless injector (Glass insert SPME, 0.8 ID; Varian, Mississauga, ON, Canada). The components were eluted on a VF-5MS capillary column, 30 m x 0.25 mm x 0.25 µm (Varian Inc., Mississuaga, ON, Canada). Compounds were detected using a Saturn 2000 mass spectrometry detector (Varian Inc., Palo Alto, CA) with a mass range of 30–400 m/z.

Volatile compounds were identified using National Institute of Standards and Technology (NIST) database (*V. 05*) as well as by comparing their mass spectra and retention times with those of the pure standards. Relative peak area, *RPA*, of each volatile compound expressed as a percentage of total volatile compounds was calculated and used to compare the flavour properties of the field peas studied.

Statistical Analysis

Analysis of variance (ANOVA) using a general linear model (GLM) procedure of the Statistical Analysis System (SAS 2004) was performed to evaluate differences between parameters. The parameters evaluated were market class, cultivar, location, crop year, replicate and interactions between them. Means comparison between parameters was carried out by Duncan's multiple range test.

RESULTS AND DISCUSSION

Effect of Market Class, Cultivar and Location on Total Volatile Compounds (TVC)

Changes in the value of TVC in different field pea cultivars grown in the year of 2008 and 2009 are, respectively, presented in Figures 1a and 1b. TVC varied significantly ($P < 0.01$) among pea cultivars (Tables 1, 4). Cooper and Rambo had the highest mean value of TVC, respectively, in the year of 2008 and 2009. Kaspa had the lowest mean value of TVC in both crop years (Tables 1, 4). Peas grown in MPK in 2008 and WIL in 2009 had higher TVC compared to other locations. On the other hand, peas from WIL in 2008 and peas from PAS in 2009 had the lowest TVC. The highest mean value of TVC was observed in peas from the marrowfat-market classes in 2008, whereas peas from the green- and dun-market classes had the highest value of TVC in 2009 (Tables 1, 4).

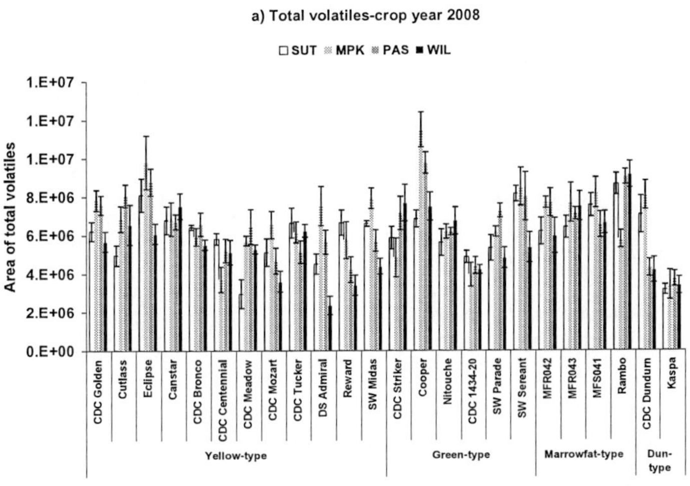

Figure 1. (Continued).

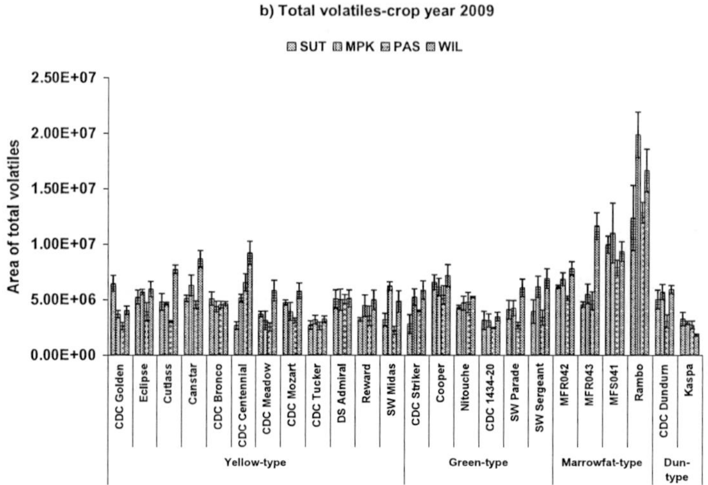

Figure 1. Changes in total volatile compounds in milled field peas as affected by market class, cultivar, and location. Results are from a two replicate analysis and expressed as mean ± standard deviation.

Effect of Market Class, Cultivar and Location on Specific Volatile Chemical Families

Alcohols

The effect of market class, cultivar, location and replication on the content of volatile alcoholic compounds in peas grown in the two different crop years is, respectively, shown in Figures 2a and 2b. The mean value of alcohols were significantly ($P < 0.01$) affected by these parameters (Tables 1, 4).

In the 2008 year, MFS041 and Nitouche had the highest mean value of alcohols, whereas Rambo had the lowest content of these compounds. Pea cultivars grown in WIL location and green-market class had higher mean value of alcohols compared to the other crops (Table 1).

In the 2009 year, CDC Striker and Rambo had, respectively, the highest and the lowest mean value of alcohols. Overall, peas grown in PAS as well as peas from dun- and green-market classes had the highest value of these compounds (Table 4).

2-Ethyl-1-hexanol was the most abundant alcohol in peas grown in both crop years (Tables 2, 5). Amongst the cultivars, SW Parade from 2008 and Rambo from 2009 had the highest mean value of this alcohol (Tables 3, 6).

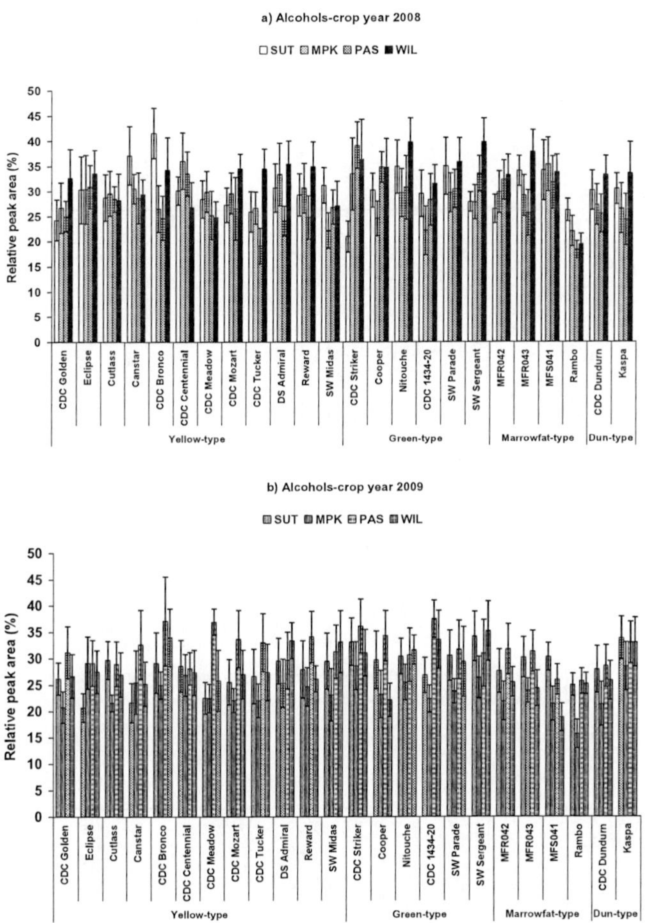

Figure 2. Changes in total alcohols in milled field peas as affected by market class, cultivar, and location. Results are from a two replicate analysis and expressed as mean ± standard deviation. Relative peak area (%) = Peak area of total alcohols/ Total peak area of volatile compounds x 100.

Alcohols in peas are mostly formed from enzymatic oxidation of lipids. Physical damage, storage and processing of seeds could lead to the formation of alcohols (Oomah and Liang 2007).

As different alcohols impart specific aromas and flavours to foods, changes in the composition of alcoholic compounds could affect the taste and flavour of peas.

Table 1. ANOVA and Duncan's multiple range test results for total volatile compounds and chemical families in milled field pea cultivars grown in the year of 2008

ANOVA[1]		Chemical family						
	Total volatile	Alcohol	Aldehyde	Ketone	Sulfur	Hydrocarbon	Pyrazine	Ester
Main effects								
	cv^2 (+++)	cv^2 (+++)	cv^2 (+++)	cv^2 (+++)	cv^2 (+++)	cv^2 (+++)	cv^2 (+++)	cv^2 (+++)
	l^3 (+++)	l^3 (+++)	l^3 (+++)	l^3 (+++)	l^3 (+++)	l^3 (+++)	l^3 (+++)	l^3 (+++)
	t^4 (+++)	t^4 (+++)	t^4 (+++)	t^4 (+++)	t^4 (+++)	t^4 (+++)	t^4 (+++)	t^4 (+++)
	r^5 (NS)	r^5 (SN)	r^5 (NS)	r^5 (NS)	r^5 (NS)	r^5 (NS)	r^5 (NS)	r^5 (++)
Interactions								
	cv*l (+++)	cv*l (+++)	cv*l (+++)	cv*l (+++)	cv*l (+++)	cv*l (+++)	cv*l (+++)	cv*l (+++)
	l*t (NS)	l*t (+++)	l*t (+++)	l*t (NS)	l*t (NS)	l*t (++)	l*t (NS)	l*t (+++)
	cv*r (NS)	cv*r (NS)	cv*r (NS)	cv*r (NS)	cv*r (NS)	cv*r (NS)	cv*r (NS)	cv*r (+++)
	l*r (NS)	l*r (NS)	l*r (NS)	l*r (NS)	l*r (NS)	l*r (NS)	l*r (NS)	l*r (NS)
	t*r (NS)	t*r (NS)	t*r (NS)	t*r (NS)	t*r (NS)	t*r (NS)	t*r (NS)	t*r (NS)
Duncan grouping								
Cultivar	Cooper (a)	MFS041 (a)	CDC Patrick (1434-20) (a)	Cooper (a)	CDC Patrick (1434-20) (a)	Kaspa (a)	CDC Golden (a)	CDC Golden (a)
	Eclipse (b)	Nitouche (a)	CDC Golden (b)	CDC Striker (ab)	CDC Golden (ab)	CDC Golden (ab)	SW Midas (b)	Eclipse (a)
	Rambo (b)	SW Parade (ab)	CDC Meadow (b)	Eclipse (b)	Kaspa (ab)	Cutlass (ab)	DS Admiral (b)	DS Admiral (b)
	SW Sergeant (c)	CDC Striker (abc)	CDC Mozart (bc)	Cutlass (bc)	Cutlass (b)	CDC Patrick (1434-20) (bc)	CDC Mozart (bc)	Kaspa (b)
	MFS041 (cd)	MFR043 (abc)	Kaspa (bc)	CDC Golden (cd)	CDC Striker (b)	CDC Mozart (bc)	Nitouche (bcd)	CDC Dundurn (c)
	MFR043 (cd)	Cooper (abcd)	DS Admiral (bcd)	CDC Patrick (1434-20) (de)	SW Midas (b)	CDC Tucker (bc)	CDC Centennial (cd)	CDC Tucker (cd)
	Canstar (de)	DS Admiral (bcde)	CDC Dundurn (cde)	DS Admiral (def)	CDC Mozart (c)	DS Admiral (bc)	CDC Dundurn (cd)	SW Midas (de)

ANOVA[1]		Chemical family						
	Total volatile	Alcohol	Aldehyde	Ketone	Sulfur	Hydrocarbon	Pyrazine	Ester
Cultivar	MFR042 (de)	Canstar (bcde)	Cutlass (def)	CDC Tucker (def)	DS Admiral (c)	CDC Striker (bc)	SW Sergeant (cd)	CDC Meadow (de)
	CDC Golden (def)	SW Sergeant (bcde)	CDC Striker (efg)	CDC Mozart (defg)	CDC Tucker (c)	SW Midas (bcd)	CDC Tucker (cd)	Cooper (de)
	Cutlass (efg)	CDC Bronco (bcde)	CDC Tucker (efgh)	MFR043 (efgh)	CDC Dundurn (cd)	Reward (bcd)	CDC Patrick (1434-20) (cde)	Rambo (d)
	CDC Striker (fgh)	CDC Centennial (bcdef)	Cooper (efghi)	Reward (efgh)	Cooper (cde)	SW Parade (bcd)	SW Parade (def)	SW Sergeant (f)
	Nitouche (gh)	Reward (bcdef)	MFR043 (efghi)	SW Parade (fgh)	CDC Centennial (cde)	Cooper (cd)	CDC Striker (efg)	MFR042 (f)
	SW Midas (gh)	MFR042 (cdefg)	Reward (fghij)	CDC Bronco (ghi)	CDC Meadow (cdef)	CDC Dundurn (de)	CDC Meadow (fg)	CDC Striker (f)
	CDC Bronco (gh)	Eclipse (cdefg)	SW Midas (fghij)	Nitouche (hij)	Reward (cdef)	CDC Meadow (e)	Canstar (g)	SW Parade (f)
	CDC Tucker (h)	CDC Dundurn (cdefg)	SW Parade (ghijk)	Rambo (hij)	Eclipse (def)	Eclipse (e)	Reward (g)	Reward (f)
	CDC Dundurn (h)	Kaspa (defg)	Rambo (ghijk)	CDC Centennial (hij)	Nitouche (efg)	CDC Centennial (e)	Kaspa (g)	Canstar (f)
	SW Parade (h)	CDC Mozart (efgh)	Nitouche (hijkl)	CDC Meadow (hij)	SW Parade (fgh)	MFS041 (ef)	MFS041 (g)	CDC Bronco (f)
	CDC Meadow (i)	CDC Patrick (1434-20) (fghi)	MFR042 (hijkl)	Kaspa (hij)	Nitouche (fghi)	MFR043 (efg)	CDC Bronco (g)	CDC Patrick (1434-20) (f)
	Reward (i)	Cutlass (ghi)	CDC Bronco (hijkl)	MFR042 (hij)	SW Sergeant (ghij)	CDC Bronco (efgh)	Cutlass (gh)	Cutlass (f)

Table 1. (Continued)

Duncan grouping								
ANOVA[1]	Chemical family							
Cultivar	Total volatile	Alcohol	Aldehyde	Ketone	Sulfur	Hydrocarbon	Pyrazine	Ester
	DS Admiral (i)	SW Midas (ghi)	MFS041 (ijkl)	SW Midas (ijk)	Canstar (hij)	Nitouche (fghi)	Eclipse (h)	MFR043 (f)
	CDC Mozart (i)	CDC Golden (ghi)	Eclipse (jklm)	MFS041 (jk)	MFR042 (ij)	SW Sergeant (ghi)	MFR043 (i)	MFS041 (f)
	CDC Centennial (i)	CDC Tucker (hi)	SW Sergeant (klm)	CDC Dundurn (jk)	MFR043 (jk)	MFR042 (ghi)	Cooper (j)	CDC Centennial (f)
	CDC Patrick (1434-20) (j)	CDC Meadow (i)	CDC Centennial (lm)	Canstar (k)	MFS041 (k)	Canstar (hi)	MFR042 (j)	Nitouche (f)
	Kaspa (k)	Rambo (j)	Canstar (m)	SW Sergeant (l)	Rambo (l)	Rambo (l)	Rambo (k)	CDC Mozart (f)
Location	Meath Park (a)	Wilkie (a)	Pasqua (a)	Wilkie (a)	Pasqua (a)	Meath Park (a)	Wilkie (a)	Meath Park (a)
	Pasqua (b)	Sutherland (b)	Meath Park (a)	Meath Park (a)	Sutherland (b)	Pasqua (a)	Meath Park (b)	Pasqua (b)
	Sutherland (c)	Meath Park (c)	Sutherland (a)	Pasqua (b)	Wilkie (b)	Wilkie (b)	Pasqua (c)	Sutherland (b)
	Wilkie (d)	Pasqua (c)	Wilkie (b)	Sutherland (c)	Meath Park (b)	Sutherland (c)	Sutherland (d)	Wilkie (c)
Market class	Marrowfat (a)	Green (a)	Dun (a)	Green (a)	Dun (a)	Dun (a)	Yellow (a)	Dun (a)
	Green (b)	Marrowfat (b)	Green (b)	Yellow (a)	Yellow (b)	Yellow (b)	Dun (a)	Yellow (b)
	Yellow (b)	Dun (b)	Yellow (b)	Marrowfat (b)	Green (b)	Green (b)	Green (a)	Green (c)
	Dun (c)	Yellow (b)	Marrowfat (c)	Dun (b)	Marrowfat (c)	Marrowfat (c)	Marrowfat (b)	Marrowfat (c)

[1]ANOVA performed using general linear model. . +++=P<0.01, ++=P<0.05, NS= Not significant (P>0.05).

[2]cv=Cultivar, [3]l=Location, t[4]=Market class, r[5]=Replicate.

Items with different letters within a column are significantly different at P<0.05 (a>b>c>d>e>f>g>h>i>j>k>l>m).

Table 2. Duncan* grouping for individual flavour compounds belonging to each chemical family found in peas-Crop year 2008

Alcohols	2-Ethyl-1-hexanol (a)	1-Hexanol (b)	1-Penten-3-ol (c)	1-Heptanol (d)	1-Octanol (e)
	R-*sec*-Butanol (f)	1-Propanol (g)	1-Butanol (h)	2-Methyl-1-propanol (i)	3-Methyl-1-butanol (j)
Aldehydes	Hexanal (a)	Butanal-3-methyl (b)	Butanal-2-methyl (b)	*Trans*-2-hexanal (c)	Benzaldehyde (d)
Hydrocarbons	Styrene (a)	Toluene (b)	Undecane (b)	Furan-2-methyl (c)	Trichloromethane (d)
	Furan,2-ethyl (e)	Ethylbenzene (f)	Xylene (g)	Benzene (g)	
Ketones	2-Butanone (a)	2-Pentanone (b)			
Sulfur compounds	Dimethyl sulfide (a)	Methanethiol (b)	Disulfide dimethyl (b)		
Esters	Ethyl acetate (a)	Methyl butyrate (b)	Butanoic acid,3-methyl-, methyl ester (b)	Hexanoic acid,methyl ester (b)	
Pyrazines	2,3-diethyl-5-methyl pyrazine (a)	Ethyl pyrazine (b)			

*Items with different letters within a row are significantly different at $P<0.05$ (a>b>c>d>e>f>g>h>i>j).

Table 3. Duncan* grouping for pea cultivars based on the most abundant volatile compounds found in peas- Crop year 2008

Most abundant flavour compounds	2-Ethyl-1-hexanol	Hexanal	2-Butanone	Dimethyl sulfide	Styrene	2,3-diethyl-5-methyl pyrazine	Ethyl acetate
Cultivars	SW Parade (a)	CDC Partrick (a)	Cooper (a)	CDC Golden (a)	SW Midas (a)	CDC Golden (a)	Eclipse (a)
	CDC Centennial (ab)	CDC Meadow (b)	Eclipse (ab)	Kaspa (a)	CDC Golden (ab)	Nitouche (b)	CDC Golden (a)
	Nitouche (ab)	CDC Mozart (b)	CDC Striker (ab)	CDC Striker (ab)	CDC Tucker (abc)	CDC Striker (c)	Kaspa (b)
	CDC Mozart (ab)	DS Admiral (bc)	CDC Golden (ab)	CDC Partrick (ab)	Kaspa (abc)	SW Sergeant (c)	DS Admiral (c)
	CDC Dundurn (abc)	CDC Golden (c)	Cutlass (b)	SW Midas (b)	DS Admiral (bcd)	Kaspa (cd)	CDC Tucker (c)
	Reward (abcd)	CDC Dundurn (c)	CDC Tucker (c)	Cutlass (b)	Cutlass (bcd)	CDC Tucker (cde)	CDC Dundurn (de)
	CDC Bronco (bcd)	CDC Tucker (d)	CDC Mozart (cd)	CDC Dundurn (c)	CDC Mozart (cde)	CDC Centennial (cde)	CDC Meadow (de)
	SW Sergeant (cde)	Kaspa (de)	DS Admiral (cd)	DS Admiral (c)	CDC Striker (de)	Cutlass (cde)	Cooper (de)
	Eclipse (def)	Reward (de)	CDC Partrick (cd)	CDC Tucker (c)	SW Parade (def)	SW Parade (de)	SW Midas (e)
	CDC Tucker (ef)	Cutlass (def)	MFR043 (cde)	Cooper (c)	Reward (def)	CDC Mozart (de)	Rambo (e)
	DS Admiral (fg)	SW Parade (def)	Kaspa (def)	CDC Mozart (c)	CDC Partrick (ef)	SW Midas (de)	MFR042 (f)
	CDC Meadow (gh)	MFR043 (defg)	SW Parade (defg)	CDC Centennial (cd)	CDC Dundurn (fg)	CDC Bronco (de)	

Most abundant flavour compounds	2-Ethyl-1-hexanol	Hexanal	2-Butanone	Dimethyl sulfide	Styrene	2,3-diethyl-5-methyl pyrazine	Ethyl acetate
	MFS041 (hi)	CDC Striker (defg)	Reward (defg)	CDC Meadow (de)	Cooper (gh)	DS Admiral (de)	
	MFR042 (hi)	SW Midas (defg)	CDC Bronco (efgh)	Reward (de)	MFR043 (gh)	CDC Dundurn (e)	
	SW Midas (hi)	CDC Bronco (efg)	Rambo (efgh)	Eclipse (def)	Eclipse (gh)	CDC Meadow (e)	
	Cooper (hij)	MFR042 (efg)	MFR042 (efghi)	Nitouche (def)	Nitouche (hi)	MFS041 (e)	
	Kaspa (ij)	Rambo (fg)	CDC Centennial (egfhi)	CDC Bronco (ef)	CDC Meadow (hi)	Eclipse (e)	
	Canstar (jk)	Nitouche (fg)	Nitouche (efghi)	SW Parade (efg)	CDC Bronco (hi)	Canstar (e)	
	Cutlass (kl)	Cooper (fg)	SW Sergeant (efghi)	SW Sergeant (fgh)	Canstar (hi)	CDC Partrick (e)	
	CDC Striker (klm)	MFS041 (fgh)	CDC Meadow (fghi)	Canstar (ghi)	MFS041 (i)	Reward (e)	
	CDC Partrick (lmn)	Canstar (ghi)	SW Midas (ghi)	MFR042 (hi)	MFR042 (i)	MFR043 (f)	
	MFR043 (lmn)	CDC Centennial (hi)	MFS041 (hi)	MFR043 (i)	SW Sergeant (i)	Cooper (f)	
	CDC Golden (mn)	SW Sergeant (hi)	CDC Dundurn (i)	MFS041 (j)	Rambo (j)	MFR042 (g)	
	Rambo (n)	Eclipse (i)	Canstar (i)	Rambo (k)	CDC Centennial (k)	Rambo (h)	

*Items with different letters within a column are significantly different at $P<0.05$
(a>b>c>d>e>f>g>h>i>j>k>l>m>n).

Table 4. ANOVA and Duncan's multiple range test results for total volatile compounds and chemical families in milled field pea cultivars grown in the year of 2009

ANOVA[1]	Chemical family							
	Total volatile	Alcohol	Aldehyde	Ketone	Sulfur	Hydrocarbon	Pyrazine	Ester
Main effects								
	cv^2 (+++)	cv^2 (+++)	cv^2 (+++)	cv^2 (+++)	cv^2 (+++)	cv^2 (+++)	cv^2 (+++)	cv^2 (+++)
	l^3 (+++)	l^3 (+++)	l^3 (+++)	l^3 (+++)	l^3 (+++)	l^3 (+++)	l^3 (+++)	l^3 (NS)
	t^4 (+++)	t^4 (+++)	t^4 (+++)	t^4 (+++)	t^4 (NS)	t^4 (+++)	t^4 (+++)	t^4 (+++)
	r^5 (NS)	r^5 (SN)	r^5 (++)	r^5 (NS)	r^5 (++)	r^5 (+++)	r^5 (NS)	r^5 (NS)
Interactions								
	cv*l (+++)	cv*l (+++)	cv*l (+++)	cv*l (+++)	cv*l (+++)	cv*l (+++)	cv*l (+++)	cv*l (+++)
	l*t (+++)	l*t (+++)	l*t (++)	l*t (NS)	l*t (NS)	l*t (+++)	l*t (+++)	l*t (+++)
	cv*r (NS)	cv*r (+++)	cv*r (NS)	cv*r (+++)	cv*r (NS)	cv*r (+++)	cv*r (+++)	cv*r (++)
	l*r (NS)	l*r (NS)	l*r (+++)	l*r (+++)	l*r (NS)	l*r (+++)	l*r (NS)	l*r (+++)
	t*r (NS)	t*r (NS)	t*r (NS)	t*r (NS)	t*r (NS)	t*r (NS)	t*r (NS)	t*r (NS)
Duncan grouping								
Cultivar	Rambo (a)	CDC Striker (a)	SW Midas (a)	CDC Patrick (1434-20) (a)	CDC Tucker (a)	CDC Tucker (a)	CDC Meadow (a)	Eclipse (a)
	MFS041 (b)	SW Sergeant (ab)	SW Parade (ab)	CDC Tucker (ab)	CDC Striker (a)	CDC Patrick (1434-20) (b)	CDC Tucker (a)	CDC Golden (b)
	MFR043 (c)	Kaspa (ab)	CDC 1434-20 (ab)	CDC Mozart (bc)	Kaspa (ab)	SW Parade (bc)	CDC Patrick (1434-20) (a)	SW Midas (c)
	MFR042 (cd)	CDC Bronco (bc)	CDC Mozart (abc)	Reward (bc)	CDC Patrick (1434-20) (bc)	Kaspa (bc)	SW Parade (b)	Reward (cd)
	Cooper (cd)	CDC Patrick (1434-20) (cd)	CDC Dundurn (abc)	CDC Meadow (bcd)	CDC Golden (c)	CDC Mozart (cd)	Kaspa (b)	CDC Meadow (cde)
	Canstar (cd)	Nitouche (cd)	CDC Tucker (bc)	SW Parade (cde)	SW Sergeant (d)	CDC Golden (cde)	SW Midas (b)	CDC Mozart (cde)
	CDC Centennial (d)	DS Admiral (cd)	CDC Golden (cd)	CDC Bronco (def)	SW Midas (d)	CDC Meadow (def)	Reward (bc)	Cooper (cde)

Duncan grouping
ANOVA[1]

| Total volatile | Chemical family | | | | | | |
	Alcohol	Aldehyde	Ketone	Sulfur	Hydrocarbon	Pyrazine	Ester
Eclipse (e)	SW Midas (cde)	DS Admiral (de)	SW Midas (efg)	Nitouche (e)	Cutlass (ef)	CDC Bronco (bcd)	MFS041 (cde)
SW Sergeant (ef)	SW Parade (cdef)	CDC Meadow (de)	Kaspa (efg)	Reward (e)	Reward (efg)	CDC Golden (bcd)	DS Admiral (cde)
Cutlass (efg)	Reward (defg)	Cutlass (e)	Eclipse (efgh)	CDC Dundurn (ef)	SW Midas (fgh)	SW Sergeant (bcde)	CDC Tucker (cdef)
DS Admiral (efg)	CDC Centennial (efgh)	CDC Bronco (ef)	Canstar (fghi)	SW Parade (ef)	MFR042 (fgh)	CDC Dundurn (bcde)	MFR043 (cdef)
CDC Dundurn (efgh)	Cooper (fgh)	CDC Centennial (efg)	MFR043 (ghij)	Cutlass (efg)	CDC Centennial (fghi)	Cutlass (bcde)	CDC Patrick (1434-20) (cdef)
Nitouche (efghi)	MFR043 (fgh)	Reward (efg)	Cooper (hijk)	CDC Bronco (fg)	CDC Dundurn (ghij)	DS Admiral (bcde)	Nitouche (ghi)cdefg
CDC Bronco (efghij)	CDC Tucker (fgh)	Kaspa (fgh)	CDC Dundurn (ijk)	CDC Meadow (g)	Nitouche (hij)	CDC Mozart (cde)	Kaspa (defg)
CDC Striker (fghij)	CDC Meadow (gh)	SW Sergeant (gh)	DS Admiral (ijk)	DS Admiral (g)	CDC Bronco (hij)	Nitouche (cde)	CDC Dundurn (defg)
CDC Mozart (ghijk)	MFR042 (ghi)	Nitouche (h)	Nitouche (ijk)	Cooper (h)	DS Admiral (ij)	CDC Striker (de)	CDC Striker (defg)
SW Parade (hijk)	Cutlass (ghi)	CDC Striker (i)	CDC Golden (ijkl)	CDC Mozart (h)	CDC Striker (ij)	CDC Centennial (de)	Reward (efg)
CDC Golden (ijk)	Eclipse (ghi)	Eclipse (ij)	CDC Centennial (ijkl)	Eclipse (h)	Eclipse (j)	Eclipse (e)	CDC Centennial (efg)
SW Midas (ijk)	CDC Mozart (ghi)	Canstar (ijk)	CDC Striker (jkl)	CDC Centennial (h)	Cooper (j)	MFR043 (f)	Canstar (efgh)
Reward (jk)	CDC Dundurn (hij)	MFR043 (jk)	Cutlass (kl)	Canstar (hi)	MFR043 (j)	MFR042 (f)	SW Sergeant (efgh)

Table 4. (Continued)

Duncan grouping								
ANOVA[1]	Chemical family							
	Total volatile	Alcohol	Aldehyde	Ketone	Sulfur	Hydrocarbon	Pyrazine	Ester
	CDC Golden (ijk)	Eclipse (ghi)	Eclipse (ij)	CDC Centennial (ijkl)	Eclipse (h)	Eclipse (j)	Eclipse (e)	CDC Centennial (efg)
	SW Midas (ijk)	CDC Mozart (ghi)	Canstar (ijk)	CDC Striker (jkl)	CDC Centennial (h)	Cooper (j)	MFR043 (f)	Canstar (efgh)
	Reward (jk)	CDC Dundurn (hij)	MFR043 (jk)	Cutlass (kl)	Canstar (hi)	MFR043 (j)	MFR042 (f)	SW Sergeant (efgh)
	CDC Meadow (k)	Canstar (hij)	Cooper (K)	MFR042 (kl)	MFR042 (hi)	SW Sergeant (k)	Cooper (g)	Cutlass (fgh)
	CDC Patrick (1434-20) (l)	CDC Golden (ij)	MFR042 (k)	SW Sergeant (l)	MFR043 (i)	Canstar (k)	MFS041 (g)	CDC Bronco (fgh)
	CDC Tucker (l)	MFS041 (jk)	MFS041 (l)	MFS041 (m)	MFS041 (j)	MFS041 (l)	Rambo (gh)	MFR042 (gh)
	Kaspa (l)	Rambo (k)	Rambo (m)	Rambo (m)	Rambo (k)	Rambo (m)	Canstar (h)	Rambo (h)
Location	Wilkie (a)	Pasqua (a)	Meath Park (a)	Pasqua (a)	Pasqua (a)	Meath Park (a)	Pasqua (a)	Sutherland (a)
	Meath Park (b)	Sutherland (b)	Sutherland (b)	Meath Park (a)	Sutherland (b)	Pasqua (a)	Sutherland (b)	Pasqua (a)
	Sutherland (c)	Wilkie (b)	Wilkie (c)	Sutherland (b)	Meath Park (c)	Sutherland (b)	Meath Park (c)	Wilkie (a)
	Pasqua (d)	Meath Park (c)	Pasqua (d)	Wilkie (c)	Wilkie (d)	Wilkie (c)	Wilkie (d)	Meath Park (a)
Market class	Green (a)	Green (a)	Dun (a)	Yellow (a)	Dun (a)	Dun (a)	Dun (a)	Yellow (a)
	Dun (a)	Dun (a)	Yellow (ab)	Green (a)	Green (a)	Yellow (a)	Yellow (a)	Green (ab)
	Yellow (b)	Yellow (b)	Green (b)	Dun (a)	Yellow (b)	Green (a)	Green (a)	Dun (ab)
	Marrowfat (c)	Marrowfat (c)	Marrowfat (c)	Marrowfat (b)	Marrowfat (c)	Marrowfat (b)	Marrowfat (b)	Marrowfat (b)

[1]ANOVA performed using general linear model. . +++=$P<0.01$, ++=$P<0.05$, NS= Not significant ($P>0.05$).

[2]cv=Cultivar, [3]l=Location, t[4]=Market class, r[5]=Replicate. Items with different letters within a column are significantly different at $P<0.05$ (a>b>c>d>e>f>g>h>i>j>k>l>m).

Table 5. Duncan* grouping for individual flavour compounds belonging to each chemical family found in peas-Crop year 2009

Alcohols	2-Ethyl-1-hexanol (a)	1-Hexanol (b)	1-Penten-3-ol (c)	1-Heptanol (d)	1-Octanol (e)
	R-*sec*-Butanol (ef)	1-Propanol (f)	2-Methyl-1-propanol (g)	1-Butanol (g)	2-Methyl-1-butanol (h)
Aldehydes	Hexanal (a)	Butanal-3-methyl (b)	Butanal-2-methyl (c)	Benzaldehyde (d)	*Trans*-2-hexanal (d)
Ketones	2-Butanone (a)	2-Pentanone (b)			
Sulfur compounds	Dimethyl sulfide (a)	Methanethiol (b)	Disulfide dimethyl (b)		
Hydrocarbons	Styrene (a)	Furan,2-ethyl (b)	Undecane (c)	Furan-2-methyl (d)	Toluene (e)
	Trichloromethane (f)	Ethylbenzene (g)	Benzene (h)	Xylene (i)	Benzene propyl (i)
Pyrazines	2,3-diethyl-5-methyl pyrazine (a)	Ethyl pyrazine (b)			
Esters	Hexanoic acid,methyl ester (a)	Butanoic acid,3-methyl-, methyl ester (b)	Ethyl acetate (c)		

*Items with different letters within a row are significantly different at *P*<0.05 (a>b>c>d>e>f>g>h>i).

Moreover the aroma threshold values (ATV) of different alcoholic compounds vary. ATVs of heptanol (3 ppb), octanol (42-480 ppb), 1-penten-3-ol (400 ppb), hexanol (200-2500 ppb), for example, are lower than other alcohols found in peas. Furthermore, taste threshold values (TTV) also varies for different alcoholic compounds. TTV of 1-octanol, 1-penten-3-ol and 1-hexanol are respectively, 2,000, 15,000 and 20,000 ppb (Burdock 2002). Differences in the composition of these compounds could, therefore, have major impacts on the aroma characteristics of peas.

Aldehydes

Changes in the content of aldehyde compounds in pea cultivars as affected by different parameters are presented in Figure 3. The mean value of aldehydes was significantly (*P* < 0.01) affected by the parameters studied

(Tables 1, 4). CDC Patrick (1434-20) had the highest and Canstar had the lowest mean value of aldehydes in 2008 (Table 1). No differences ($P > 0.05$) were observed in the mean value of aldehydes in the peas grown in PAS, MPK and SUT in this crop year. Dun-market class contained the highest and marrowfat-market class had the lowest mean value of these carbonyl compounds (Table 1).

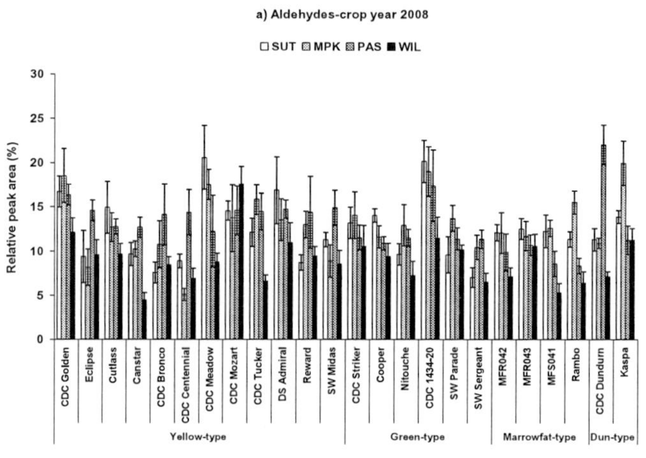

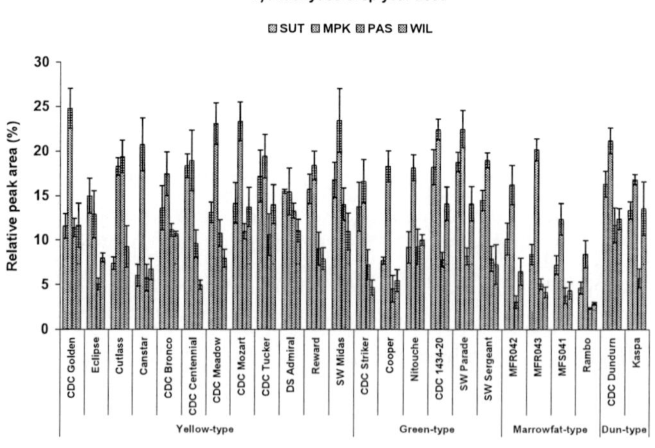

Figure 3. Changes in total aldehydes in milled field peas as affected by market class, cultivar, and location. Results are from a two replicate analysis and expressed as mean ± standard deviation. Relative peak area (%) = Peak area of total aldehydes/ Total peak area of volatile compounds x 100.

In the 2009 crop year, SW Midas and Rambo had, respectively, the highest and the lowest levels of these compounds. Furthermore, cultivars grown in MPK as well as dun-market class had the highest mean value of aldehydes (Table 4).

Hexanal was the most abundant aldehyde found in the pea cultivars (Tables 2, 5). CDC Patrick (1434-20) and SW Midas, respectively, from 2008 and 2009 were the cultivars with the highest abundance of this compound (Tables 3, 6).

Enzymatic or autoxidative decomposition of unsaturated fatty acids, mainly linoleic and linolenic acids could lead to the formation of aldehydes in peas (Hornostaj and Robinson 2000; Barra et al. 2007). Differences observed in the concentration of these carbonyl compounds could be due to differences in linoleate compositions in pea cultivars (Oomah and Liang 2007).

Hexanal is commonly identified in fruits and vegetables (Hornostaj and Robinson 2000; Oomah and Liang 2007) and is reportedly responsible for the off-flavour in stored unblanched frozen peas (Barra et al. 2007). It has a fatty, green, grassy, fruity odour and taste (Burdock 2002). Hexanal and *trans*-2-hexenal have lower ATV (respectively, 4.1-22.8 ppb and 30 ppb) than the other aldehydes found in the peas studied. Thus, they could dominate the aroma and flavour of peas. Additionally, *trans*-2-hexenal and butanal-2-methyl, respectively, have TTVs of 10,000 and 30,000 ppb (Burdock 2002) and could also influence the taste of peas. Timely harvesting and prevention of physical damages during operation may prevent the formation of undesirable aldehyde flavours in peas (Hornostaj and Robinson 2000).

Ketones

Changes in the amount of ketones determined in peas are shown in Figure 4. Differences ($P < 0.01$) in the mean value of ketones were observed between pea cultivars (Tables 1, 4). In the 2008 year, the highest mean value of ketones was found in the Cooper cultivar, whereas SW Sergeant had the lowest value of this chemical group. Pea cultivars grown in WIL and MPK contained the highest amount of ketones and those grown in SUT had the lowest mean value. Green- and yellow-market classes contained higher mean values of ketones than the other market classes (Table 1). In the 2009 crop year, the highest mean value of ketones was found in the CDC Patrick (1434-20) cultivar, whereas MFS041 and Rambo had the lowest value of ketones. Peas grown in PAS and MPK had higher contents of ketones than those grown in the other locations.

Yellow-, green-, and dun-market classes had the same level of this chemical group (Table 4).

2-Butanone was present in higher amounts compared to 2-pentanone in the pea cultivars (Tables 2, 5). Amongst the pea cultivars, Cooper from 2008 and CDC Tucker from 2009 had the greatest amount of this compound (Tables 3, 6).

Ketones are derived from lipid oxidation. 2-Pentanone, and 2-butanone have been described as having a wine or acetone odour, and a sweet apricot odour, respectively (Burdock 2002). 2-Pentanone has ATV of 70 ppb and TTV of 25,000 ppb, whereas TTV of 2-butanone is 5,000 ppb.

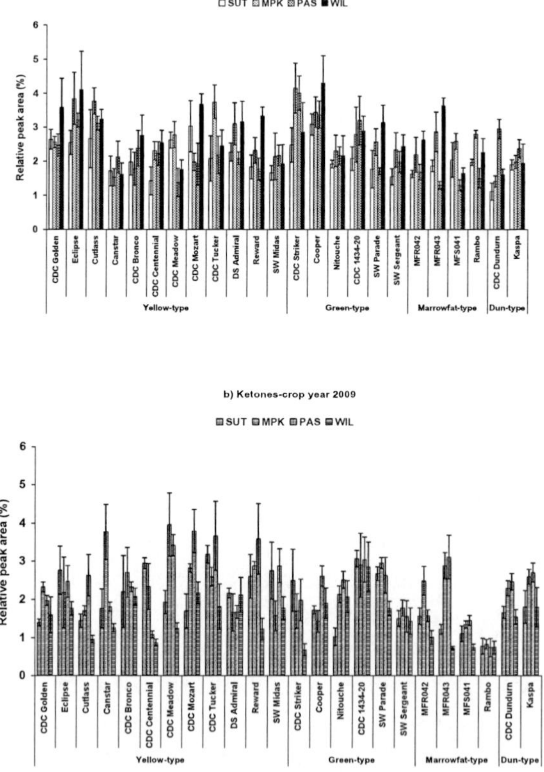

Figure 4. Changes in total ketones in milled field peas as affected by market class, cultivar, and location. Results are from a two replicate analysis and expressed as mean ± standard deviation. Relative peak area (%) = Peak area of total ketones/ Total peak area of volatile compounds x 100.

Sulfur Compounds

Differences in the content of sulfur compounds between the pea cultivars are presented in Figure 5. In the 2008 year, the highest mean value of sulfur compounds was observed in CDC Patrick (1434-20), and the lowest was found in the Rambo cultivar. Pea cultivars grown in PAS had higher mean value of sulfur containing compounds than those grown in the other locations. Dun- and marrowfat-market classes had, respectively, the highest and the lowest level of sulfur containing compounds. No differences ($P > 0.05$) were found between yellow- and green-market classes (Table 1). In the crop year of 2009, the highest mean value of these compounds was observed in the CDC Tucker and CDC Striker cultivars, and the lowest was found in the Rambo cultivar.

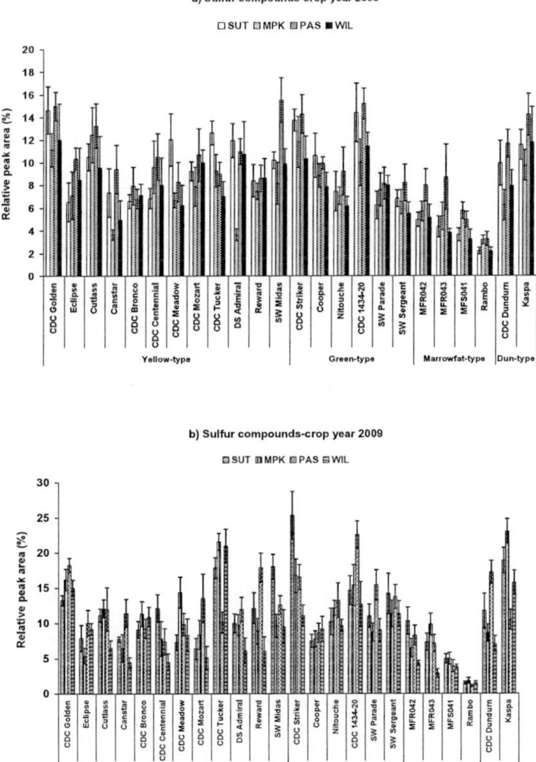

Figure 5. Changes in total sulfur compounds in milled field peas as affected by market class, cultivar, and location.Results are from a two replicate analysis and expressed as mean ± standard deviation. Relative peak area (%) = Peak area of total sulfur compounds/ Total peak area of volatile compounds x 100.

The highest sulfur compounds were observed in peas from PAS, whereas peas from WIL had the lowest value of these compounds. Peas from dun- and green-market classes had higher sulfur containing compounds than the other market classes (Table 4). Volatile sulfur compounds are natural compounds in foods and could be formed during heat processing and storage (Maga 1982). Formation of these compounds has been reported in blanched peas (Jakobsen et al. 1998). Sulfur compounds contribute to the overall flavour and aroma of foods (Jakobsen et al., 1998). For example, dimethyl disulfide has a diffuse, intense onion odour with an ATV of 0.16-1.2 ppb. Dimethyl sulfide, on the other hand, has an intense, cabbage odour with an ATV of 0.3-10 ppb (Burdock 2002). These sulfur compounds with such low ATVs could markedly impact the aroma of peas. Dimethyl sulfide was the most abundant sulfur compound in the peas (Tables 2, 5). CDC Golden and Kaspa cultivars from 2008 and CDC Tucker and CDC Striker cultivars from 2009 had the highest level of this sulfur compound (Tables 3, 6).

Hydrocarbons

Significant ($P < 0.01$) differences in the mean value of hydrocarbons were observed between the pea cultivars grown in the year of 2008 and 2009 (Figure 6, Tables 1,4). In the 2008 year, the highest and the lowest mean value of hydrocarbons were, respectively, found in the Kaspa and Rambo cultivars. Peas grown in MPK and PAS had the highest value of this chemical family, whereas peas grown in SUT had the lowest value of these compounds.

Dun- and marrowfat-market classes, respectively, had the highest and the lowest value of hydrocarbons. No significant differences were observed between yellow- and green-market classes (Table 1).

In the crop year of 2009, CDC Tucker and Rambo had, respectively, the highest and the lowest mean value of this chemical group (Table 4). Peas grown in MPK and PAS had the highest mean value of hydrocarbons. No differences were found in these compounds in dun-, yellow-and green-market classes (Table 4).

The most abundant hydrocarbon was styrene (Tables 2, 5) which has a sweet, balsamic, floral odour. This compound has an ATV of 3.6-80 ppb and was dominant in the SW Midas, CDC Golden and CDC Tucker cultivars (Tables 3, 6).

In general, hydrocarbons are derived from oxidation of unsaturated fatty acids in foods (Märk et al. 2006). Trichloromethane (chloroform) is a natural compound in plants (Lovegren et al. 1979). Volatile alkanes reportedly contribute to desirable odour or flavour characteristics of green beans and peas

(Perkins 1989). Furan,2-ethyl has a warm, sweet odour and a coffee-like flavour (Burdock 2002). These compounds can therefore play a contributory role in the flavour profile of peas.

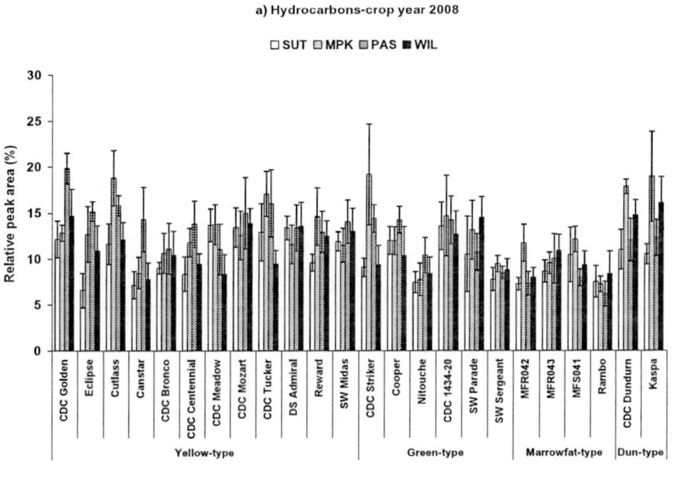

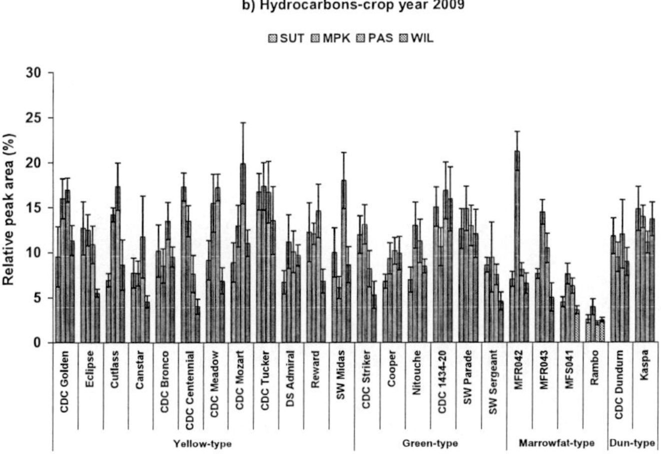

Figure 6. Changes in total hydrocarbons in milled field peas as affected by market class, cultivar, and location. Results are from a two replicate analysis and expressed as mean ± standard deviation. Relative peak area (%) = Peak area of total hydrocarbons/ Total peak area of volatile compounds x 100.

Pyrazines

Changes in the content of pyrazine containing volatile compounds in the peas are shown in Figure 7. ANOVA results showed significant ($P < 0.01$) differences in the pyrazines found in peas (Tables 1, 4).Amongst the cultivars grown in the year of 2008, CDC Golden and Rambo had, respectively, the highest and the lowest value of this chemical group. In general, peas grown in WIL had higher content of pyrazines than the other crops. No significant differences in this chemical group were found between the yellow-, green-, and dun-market classes (Table 1).

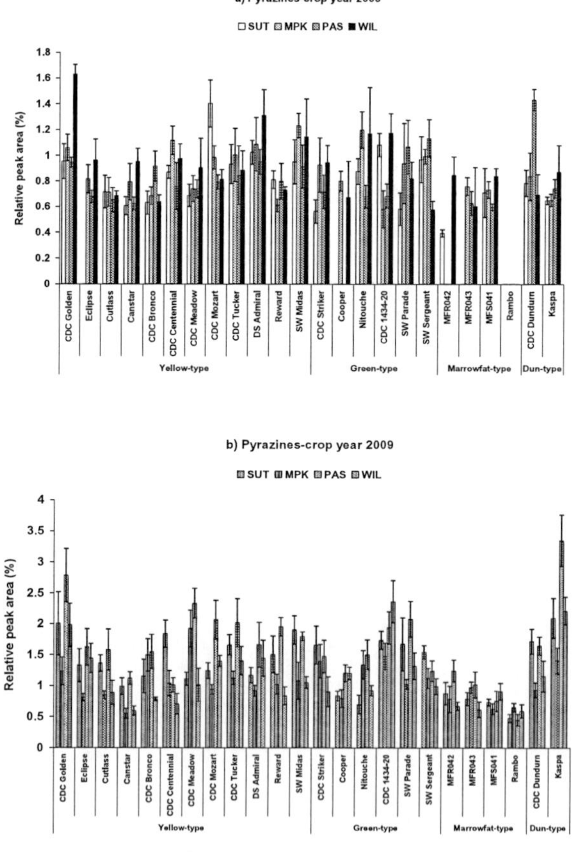

Figure 7. Changes in total pyrazines in milled field peas as affected by market class, cultivar, and location. Results are from a two replicate analysis and expressed as mean ± standard deviation. Relative peak area (%) = Peak area of total pyrazines/ Total peak area of volatile compounds x 100.

In the 2009 year, CDC Meadow, CDC Tucker and CDC Patrick (1434-20) cultivars were found to have the highest value of pyrazines, whereas the Canstar cultivar had the lowest value of this chemical group (Table 4). Furthermore, peas grown in PAS had higher values of these compounds than those grown in the other locations. Duncan's test showed no differences ($P >$ 0.05) between dun-, green-, and yellow-market classes; the lowest value of this chemical group was observed in marrowfat-market class (Table 4).

2,3-Diethyl-5-methyl pyrazine was the most prominent pyrazine identified in peas (Tables 2, 5). This compound has an ATV of 0.09-1 ppb (Burdock, 2002). CDC Golden and Kaspa had greater amounts of this compound than the other crops (Tables 3, 6).

Pyrazines have low vapour pressure and intense smell. These compounds contribute to desirable flavour and aroma of fresh vegetables (Müller and Rappert 2010), and have been reported in peas, beans and soybeans (Burdock 2002). 2,3-Diethyl-5-methyl pyrazine has a nutty, meaty, roasted hazelnut odour, whereas ethyl pyrazine has a peanut butter, musty, nutty, woody, buttery odour (Burdock 2002).

Esters

The content of esteric compounds was affected by the market class, cultivar and crop year (Figure 8, Tables 1, 4). No differences ($P >$ 0.05) were found between the cultivars grown in different locations in the year of 2009 (Table 4).

In the 2008 crop year, CDC Golden and Eclipse cultivars had the highest value of esters. Cultivars grown in MPK as well as dun-market class had the highest value of esters (Table 1).In 2009, Eclipse and Rambo had, respectively, the highest and the lowest mean value of esters. In general, the yellow-market class had higher mean value of esters than the other market classes (Table 4). Hexanoic acid,methyl ester and ethyl acetate were the most abundant esters in the peas (Tables 2, 5). ATVs of hexanoic acid,methyl ester and ethyl acetate are, respectively, 10-87 ppb and 5-5,000 ppb, whereas this value for butanoic acid,3-methyl-, methyl ester is 4.4-44 ppb (Burdock 2002).

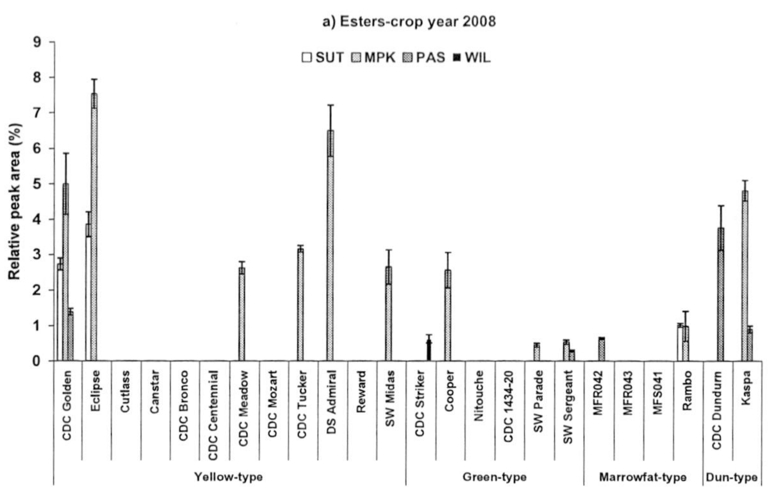

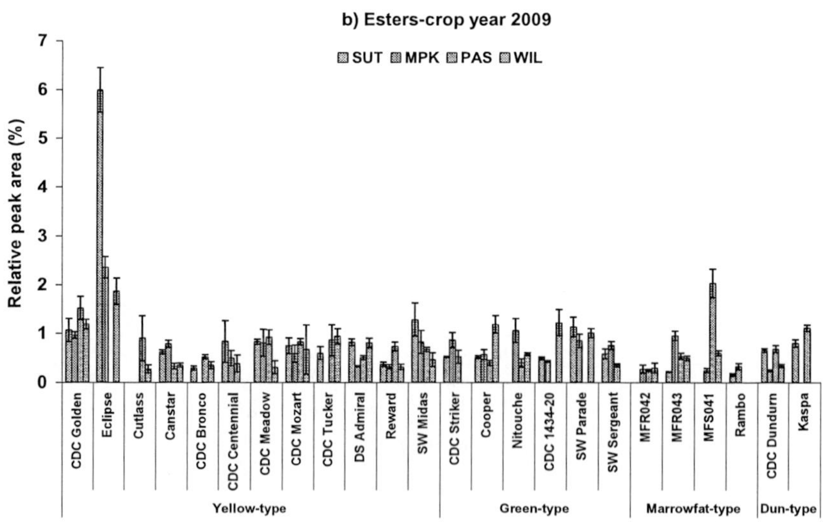

Figure 8. Changes in total esters in milled field peas as affected by market class, cultivar, and location. Results are from a two replicate analysis and expressed as mean ± standard deviation. Relative peak area (%) = Peak area of total esters/ Total peak area of volatile compounds x 100.

Table 6. Duncan* grouping for pea cultivars based on the most abundant volatile compounds found in peas-Crop year 2009

Most abundant flavour compounds	2-Ethyl-1-hexanol	Hexanal	2-Butanone	Dimethyl sulfide	Styrene	2,3-diethyl-5-methyl pyrazine	Hexanoic acid,methyl ester
Cultivars	Rambo (a)	SW Midas (a)	CDC Tucker (a)	CDC Tucker (a)	CDC Tucker (a)	Kaspa (a)	CDC Meadow (a)
	MFS041 (b)	SW Parade (ab)	CDC Partrick (b)	CDC Striker (a)	CDC Golden (a)	CDC Golden (a)	CDC Mozart (a)
	MFR042 (c)	CDC Partrick (abc)	Reward (bc)	Kaspa (b)	Cutlass (b)	CDC Partrick (b)	CDC Golden (b)
	MFR043 (cd)	CDC Mozart (abc)	CDC Golden (bcd)	CDC Partrick (b)	CDC Partrick (b)	SW Parade (c)	CDC Tucker (cb)
	Canstar (de)	CDC Dundurn (abc)	CDC Mozart (bcd)	CDC Golden (b)	Reward (bc)	CDC Meadow (c)	Kaspa (bcd)
	Cooper (de)	CDC Tucker (bc)	Kaspa (bcde)	SW Sergeant (c)	CDC Centennial (cd)	Eclipse (cd)	CDC Dundurn (bcd)
	Eclipse (e)	CDC Golden (cd)	CDC Bronco (cdef)	SW Midas (c)	SW Parade (de)	CDC Tucker (cd)	CDC Striker (bcd)
	Cutlass (ef)	Cutlass (de)	SW Midas (cdef)	Reward (d)	Kaspa (def)	CDC Dundurn (cde)	DS Admiral (cd)
	CDC Centennial (fg)	DS Admiral (de)	CDC Meadow (cdef)	Nitouche (d)	CDC Mozart (def)	CDC Mozart (cde)	Reward (cd)
	CDC Bronco (gh)	CDC Bronco (e)	SW Parade (cdef)	CDC Dundurn (de)	CDC Meadow (def)	DS Admiral (cde)	CDC Centennial (cd)
	DS Admiral (ghi)	CDC Centennial (ef)	Eclipse (defg)	Cutlass (def)	CDC Dundurn (ef)	CDC Striker (cde)	CDC Partrick (cd)

Table 6. (Continued)

Most abundant flavour compounds	2-Ethyl-1-hexanol	Hexanal	2-Butanone	Dimethyl sulfide	Styrene	2,3-diethyl-5-methyl pyrazine	Hexanoic acid,methyl ester
	SW Midas (hij)	Reward (ef)	DS Admiral (efgh)	SW Parade (def)	Eclipse (ef)	SW Midas (cde)	Cooper (d)
	Reward (ijk)	CDC Meadow (ef)	MFR043 (fgh)	CDC Bronco (ef)	Cooper (ef)	Nitouche (def)	SW Parade (d)
	CDC Striker (jk)	SW Sergeant (fg)	CDC Dundurn (gh)	CDC Meadow (fg)	SW Sergeant (efg)	Reward (ef)	SW Sergeant (e)
	SW Sergeant (jk)	Kaspa (g)	SW Sergeant (gh)	DS Admiral (fg)	MFR043 (efg)	SW Sergeant (ef)	MFR043 (ef)
	Nitouche (jk)	CDC Striker (h)	Cooper (gh)	Cooper (gh)	CDC Striker (efgh)	CDC Bronco (fg)	Cutlass (ef)
	CDC Dundurn (k)	Canstar (h)	Nitouche (gh)	Eclipse (hi)	CDC Bronco (fgh)	CDC Centennial (fgh)	CDC Bronco (ef)
	CDC Golden (l)	Nitouche (hi)	CDC Striker (hi)	CDC Mozart (hi)	Nitouche (ghi)	Cooper (ghi)	SW Midas (ef)
	CDC Meadow (lm)	MFR043 (hi)	Cutlass (hi)	CDC Centennial (hi)	DS Admiral (hi)	Canstar (ghi)	Canstar (ef)
	CDC Tucker (mn)	Eclipse (ij)	CDC Centennial (hi)	Canstar (hij)	Canstar (hi)	MFR042 (ghi)	Eclipse (ef)
	SW Parade (mn)	MFR042 (j)	Canstar (i)	MFR042 (ij)	MFR042 (i)	Cutlass (hi)	Nitouche (f)
	Kaspa (mn)	Cooper (j)	MFR042 (i)	MFR043 (j)	SW Midas (i)	MFS041 (ij)	MFR042 (fg)
	CDC Mozart (mn)	MFS041 (k)	MFS041 (j)	MFS041 (k)	MFS041 (i)	MFR043 (ij)	Rambo (gh)
	CDC Partrick (n)	Rambo (l)	Rambo (k)	Rambo (l)	Rambo (j)	Rambo (j)	MFS041 (h)

*Items with different letters within a column are significantly different at $P<0.05$ (a>b>c>d>e>f>g>h>i>j>k>l>m>n).

Although the mean value determined of the latter ester was lower than those of other esters, it could still have a notable effect on the aroma of peas. Eclipse and CDC Golden cultivars from 2008 and CDC Meadow and CDC Mozart cultivars from 2009 had, respectively, higher values of ethyl acetate and hexanoic acid,methyl ester compared to the other cultivars (Tables 3, 6).Ethyl acetate has an ether, brandy odour and a fruity, sweet taste, and it has been reported in soybeans and beans (Burdock 2002; del Rosario et al. 1984). Hexanoic acid,methyl ester, on the other hand, has an ether, pineapple odour (Burdock 2002), and butanoic acid,3-methyl-,methyl ester has a bitter flavour and an herbaceous, fruity odour.

Terpenes

α-Pinene was the only terpene identified in the peas, and its amount significantly ($P < 0.01$) varied among the pea cultivars. In the 2008 year, peas grown in PAS as well as marrowfat market class contained the highest amount of α-pinene, whereas peas from MPK had the lowest amount of this compound. No differences were observed between yellow and dun market classes in this compound. In the 2009 year, on the other hand, peas from PAS and WIL contained, respectively, the highest and the lowest amount of this compound. No differences were observed between peas from dun-, yellow-, and green-market class. In general, peas from the year of 2009 had higher α-pinene than those from the 2008 year. α-Pinene has a pine- and turpentine-like odour with an ATV of 2.5-62 ppb (Burdock 2002). In general, terpenes are found in essential oils and headspace of flowers and leaves (Jakobsen et al. 1998; de Schutter et al. 2008), and their presence in legumes could be due to the enzymatic degradation of carotenoid compounds (Jakobsen et al. 1998).

Effect of Crop Year on the Flavour Profile of Field Pea Cultivars

The effect of crop year on the flavour profile of pea cultivars was evaluated on the data pooled from the two crop years. Results from ANOVA analysis showed that TVC in pea was significantly ($P < 0.01$) affected by the crop year (Table 7). Cultivars grown in the year 2008 had higher TVC than those from the 2009 (Table 7). Alcohols, ketones, pyrazine and hydrocarbons were higher in the peas grown in the 2008 year. On the other hand, peas from the year 2009 had higher mean values of aldehydes, sulfur compounds, esters, and terpenes (Table 7). Peas from Marrowfat-market class as well as peas grown in WIL and MPK locations had the highest mean value of total volatiles. Amongst the cultivars, Rambo and Kaspa had, respectively, the highest and the lowest level of TVC.

Table 7. ANOVA and Duncan's multiple range test results for total volatile compounds and chemical families in milled field peas as affected by crop year

ANOVA[1]

Main effects	Interactions									
	cv[2]	l[3]	t[4]	cy[5]	r[6]	cv*l	cv*cy	cy*l	t*l	t*cy
Total volatiles	+++	+++	+++	+++	NS	+++	+++	+++	++	+++
Alcohol	+++	+++	+++	+++	NS	+++	+++	+++	+++	+++
Aldehyde	+++	+++	+++	+++	NS	+++	+++	+++	+++	+++
Ketone	+++	+++	+++	+++	NS	+++	+++	+++	++	+++
Sulfur compound	+++	+++	+++	+++	NS	+++	+++	+++	NS	+++
Hydrocarbon	+++	+++	+++	+++	NS	+++	+++	+++	+++	NS
Ester	+++	+++	+++	++	NS	+++	+++	+++	+++	+++
Pyrazine	+++	+++	+++	+++	NS	+++	+++	+++	NS	+++
Terpene	+++	+++	NS	+++	NS	+++	+++	+++	++	+++

Duncan grouping

Compound	Crop year		Market class				Location			
Alcohol	2008 (a)	2009 (b)	Green (a)	Dun (b)	Yellow (bc)	Marrowfat (c)	WIL[7] (a)	PAS[8] (a)	SUT[9] (b)	MPK[10] (c)
Aldehyde	2009 (a)	2008 (b)	Dun (a)	Yellow (b)	Green (b)	Marrowfat (c)	MPK (a)	SUT (b)	PAS (c)	WIL (d)
Ketone	2008 (a)	2009 (b)	Green (a)	Yellow (a)	Dun (b)	Marrowfat (c)	MPK (a)	PAS (b)	WIL (c)	SUT (c)
Sulfur compound	2009 (a)	2008 (b)	Dun (a)	Green (b)	Yellow (c)	Marrowfat (d)	PAS (a)	SUT (b)	MPK (c)	WIL (d)
Hydrocarbon	2008 (a)	2009 (b)	Dun (a)	Yellow (b)	Green (b)	Marrowfat (c)	MPK (a)	PAS (a)	SUT (b)	WIL (c)
Ester	2009 (a)	2008 (b)	Dun (a)	Yellow (a)	Green (b)	Marrowfat (b)	MPK (a)	SUT (b)	PAS (bc)	WIL (c)
Pyrazine	2008 (a)	2009 (b)	Yellow (a)	Dun (a)	Green (a)	Marrowfat (b)	WIL (a)	PAS (a)	MPK (a)	SUT (b)
Terpene	2009 (a)	2008 (b)	Green (a)	Yellow (a)	Dun (a)	Marrowfat (a)	PAS (a)	SUT (b)	MPK (c)	WIL (c)
Total volatiles	2008 (a)	2009 (b)	Marrowfat (a)	Green (b)	Yellow (b)	Dun (c)	WIL[7] (a)	MPK (a)	SUT (b)	PAS (b)

[1]ANOVA performed using general linear model. +++=$P<0.01$, ++= $P<0.05$, NS= Not significant ($P>0.05$).

[2]cv=Cultivar, [3]l=Location, t[4]=Market class, cy[5]=Crop year, r[6]=Replicate,

WIL[7]=Wilkie, PAS[8] =Pasqua, SUT[9]=Sutherland, MPK[10]=Meath Park. Compounds belonging to each chemical family with different letters within a row are significantly different at $P<0.05$ (a>b>c>d).

In summary, 2-Ethyl-1-hexanol, hexanal, 2-butanone, dimethyl sulfide, styrene, 2,3-diethyl-5-methyl pyrazine, ethyl acetate and hexanoic acid,methyl ester were the most abundant flavour compounds found in the field peas studied. Overall, there were significant differences in the content of the different volatile chemical families between the cultivars (Table 7). The lowest value of alcohols, aldehydes, ketones, sulphur compounds, hydrocarbons and pyrazines was found in the marrowfat-market class. In contrast, the dun-market class had the highest value of these compounds except for ketones and alcohols. Ketones were found in the highest amount in yellow- and green-market classes, and alcohols were observed at the highest level in green-market class (Table 7). Peas grown at MPK had the highest value of aldehydes, ketones and esters. In contrast, peas grown at PAS had higher sulfur containing compounds and terpenes compared to those from other locations. The highest hydrocarbons were found in pea from PAS and MPK, and the greatest level of alcohols was observed in peas grown at WIL and PAS (Table 7).

CONCLUSION

The results clearly show that the flavour profile of pea is affected by market class, cultivar, location and crop year. This work was done on raw flours and confirms some of the previous work done in our laboratory which focused on differences in the flavour properties of cooked peas (Azarnia et al. 2011). It also demonstrates some of the differences in the flavour of raw and cooked seeds. For example, the cooked Rambo from 2008 and Cutlass from 2009 had the highest mean value of alcohols, whereas raw CDC Striker from both years had the highest value of these compounds. The greatest mean value of aldehydes was observed in the cooked CDC Striker and CDC Patrick from 2008 and MFR042 from 2009. On the other hand, raw CDC Patrick from both years had the highest value of this group (Azarnia et al., 2011). Cooked CDC Dundurn, Cooper and MFR042 had the highest value of ketones in both years, whereas raw Cooper from 2008 and CDC Patrick from 2009 had the highest value of this group. As peas having different flavour profiles can affect the taste and flavour of the finished food product, knowledge of the effect of different production parameters on pea flavour could be useful for the selection of the right cultivars for different food applications. Additionally, further studies conducted over several years will be useful in determining the effect of environmental impacts on the flavour of peas.

ACKNOWLEDGMENTS

This research was funded by the Saskatchewan Pulse Growers Association and Agriculture and Agri-Food Canada.

REFERENCES

AAFC (2006). Dry Peas: Situation and Outlook. Bi-Weekly Bulletin 19 (2). Agriculture and Agri-Food Canada: Ottawa.

Azarnia, S., Boye, J. I., Warkentin, T., and Malcolmson, L. (2011). Application of gas chromatography in the analysis of flavour compounds in field peas. In B. Salih (Ed.), Gas Chromatography / Book 2. InTech-Open Access Publisher (*In press*).

Azarnia, S., Boye, J. I., Warkentin, T., Malcolmson, L., Sabik, H., and Bellido, A. S. (2010). Volatile flavour profile changes in selected field pea cultivars as affected by crop year and processing. *Food Chemistry*, 124 (1), 326-335.

Barra, A., Baldovini, N., Loiseau, A. M., Albino, L., Lesecq, C., and Lizzani Cuvelier, L. (2007). Chemical analysis of French beans (Phaseolus vulgaris L.) by headspace solid phase microextraction (HS-SPME) and simultaneous distillation/extraction (SDE). *Food Chemistry*, 101 (3), 1279-1284.

Burdock, G. A. (2002). Handbook of flavour ingredients. Boca Raton: CRC PRESS.

de Almeida Costa, G. E., da Silva Queiroz-Monici, K., Pissini Machado Reis, S. M., and de Oliveira, A. C. (2006). Chemical composition, dietary fibre and resistant starch contents of raw and cooked pea, common bean, chickpea and lentil legumes. *Food Chemistry*, 94 (3), 327–330.

de Schutter, D. P., Saison, D., Delvaux, F., Derdelinckx, G., Rock, J. M., Neven, H., and Delvaux, F. R. (2008). Release and evaporation of volatiles during boiling of unhopped wort. *Journal of Agricultural and Food Chemistry*, 56 (13), 5172-5180.

del Rosario, R., de Lumen, B. O., Habu, T., Flath, R. A., Mon, T. R., and Teranishi, R. (1984). Comparison of headspace of volatiles from winged beans and soybeans. *Journal of Agricultural and Food Chemistry*, 32 (5), 1011-1015.

Hornostaj, A. R., and Robinson, D. S. (2000). Purification of hydroperoxide lyase from pea seeds. *Food Chemistry*, 71 (2), 241-247.

Jakobsen, H. B., Hansen, M., Christensen, M. R., Brockhoff, P. B., and Olsen, C. E. (1998). Aroma volatiles of blanched green peas (Pisum sativum L.). *Journal of Agricultural and Food Chemistry*, 46 (9), 3727–3734.

Lovegren, N. V., Fisher, G. S., Legendre, M. G., and Schuller, W. H. (1979). Volatile constituents of dried legumes. *Journal of Agricultural and Food Chemistry*, 27 (4), 851-853.

Maga, J. A. (1982). Pyrazines in foods. *Critical Reviews in Food Science and Nutrition*, 16 (1), 1-48.

Märk, J., Pollien, P., Lindinger, C., Blank, I., and Märk, T. (2006). Quantitation of furan and methylfuran formed in different precursor systems by proton transfer reaction mass spectrometry. *Journal of Agricultural and Food Chemistry*, 54 (7), 2786-2793.

Müller, R., and Rappert, S. (2010). Pyrazines: Occurrence, formation and biodegradation. *Applied Microbiology and Biotechnology*, 85 (5), 1315–1320.

Oomah, B. D., and Liang, L. S. Y. (2007). Volatile compounds of dry beans (*Phaseolus vulgaris L.*). *Plant Foods for Human Nutrition*, 62 (4), 177-183.

Perkins, E. G. (1989). Gas chromatography and gas chromatography-mass spectrometry of odor and flavor components in lipid foods. In D. B. Min, and T. H. Smouse (Eds.), Flavor chemistry of lipid foods (pp. 35-56). Champaign, IL: American Oil Chemists' Society.

SAS (2004). SAS user's guide: Statistics, Version 9.1. Cary: SAS Institute.

Sathe, S. K. (2002). Dry bean protein functionality. *Critical Reviews in Biotechnology*, 22 (2), 175-223.

Tharanathan, R. N., and Mahadevamma, S. (2003). Grain legumes: A boon to human nutrition. *Trends in Food Science and Technology*, 14 (12), 507-518.

In: Peas   ISBN: 978-1-61942-866-9
Editors: A. Comstock and B. Lothrop   © 2012 Nova Science Publishers, Inc.

Chapter 3

REGULATION OF NODULE DEVELOPMENT IN *PISUM SATIVUM* L. HAS COMMON ASPECTS WITH MECHANISMS OPERATING IN REGULAR APICAL MERISTEMS

M. A. Osipova[1,2], *L. A. Lutova*[2] *and E. A. Dolgikh*[1]

[1]All-Russia Research Institute for Agricultural Microbiology, Podbelsky chausse 3, 196608, St.-Petersburg, Pushkin 8, Russia
[2]Department of Genetics, St. Petersburg State University, 7/9, University emb., 199034, St. Petersburg, Russia

ABSTRACT

The symbiotic interaction of legume plants with soil bacteria of the genus *Rhizobium* results in the formation of new organs on legume roots – nodules, where symbiotic nitrogen fixation takes place. Symbiotic nodules are formed as a result of dedifferentiation and reactivation of cell division of cortical root cells. A few data are available regarding mechanisms regulating nodule meristem development and function. In our work we focused on meristem development of indeterminate-type nodules in *Pisum sativum* L. According to recent data there are common features in regulation of nodule meristem and regular apical meristems. In this chapter we will briefly review known mechanisms underlying plant apical meristems development and maintenance with special attention to homeodomain transcription factor of WOX family and CLAVATA-related receptor complexes. Based on analysis of pea nodule development we will discuss the role of WOX and CLAVATA-related genes in nodule organogenesis.

1. INTRODUCTION

Pea (*Pisum sativum* L.) happened to be a perfect genetic model that allowed Gregor Mendel to discover the laws of inheritance and to give rise to a new age in biology - to genetic science. Now pea is still actual as a model object in genetic research due to its amazing peculiarity – the ability to form symbiotic nodules, where nitrogen fixation takes place. Being a legume plant pea is involved in symbiotic interaction with soil bacteria of the genus *Rhizobium* resulting in the formation of new specialized organs on its roots – nodules. A multitude of pea mutants having various defects in symbiosis allows discovering plant genes involved in the control of different stages of nodule development. Analysis of more than 100 pea symbiotic mutants identified more than 40 loci involved in the nodulation process (Borisov et al., 2000; Tsyganov et al., 2002). Nodules represent facultative organs that are formed only under the nitrogen-poor conditions. In this regard, the process of symbiotic nodule formation is regulated both at the level of the whole plant, i.e. systemically, and by local mechanisms. Host systemic control of nodulation is known as autoregulation of nodulation (AON). AON defines the process by which early nodulation events inhibit further nodule development on the entire root system via currently unknown long-distance signals to maintain a balance between shoot and root development (Kosslak and Bohlool, 1984). Pea mutants with overabundant number of nodules (so called «supernodulating mutants» *Pssym29, Pssym28, Psnod3*) defected in AON have been described as well as mutants of model legumes (*Lotus japonicas – Ljhar1, Ljclv2*; *Medicago truncatula – Mtsunn, Mtrdn1*) defective in orthologous genes. These mutants represent suitable model to study systemic control of nodule formation. Symbiotic nodules are formed as a result of dedifferentiation and reactivation of cortical root cells. The peculiarity of pea nodules is that they are of indeterminate type having their own persistent meristem – a population of undifferentiated proliferating cells. Mechanisms regulating nodule meristem development are still far from understanding. Progress towards the elucidation of such mechanisms has been done, when the cloning of pea *PsSym29* and *PsSym28* genes showed that they encode CLV1-like and CLV2 receptors similar to the Arabidopsis (*Arabidopsis thaliana*) receptors, regulating development of shoot apical meristem (SAM) (Krusell et al., 2002; 2011). It suggests that in pea the CLV receptors are involved in AON. Hence a growing body of evidence suggests common mechanisms regulating different types of plant meristems. The homeodomain transcription factors of WUSCHEL - RELATED HOMEOBOX (WOX) family are proposed to be the important regulators of apical meristem maintenace, being expressed in «organizer» of apical meristems (Sarkar et al., 2007). Among WOX transcription factors the WUSCHEL (WUS) regulates stem cells maintenance in shoot apical meristem and is involved in interaction with CLAVATA (CLV) system, including CLE-peptide CLV3 and CLV receptor complex, whereas its paralogue – WOX5 - is supposed to have the same function in root apical meristem. The CLV signaling pathway negatively regulates the *WUS* expression in the SAM and the same type of network may be operating in the RAM but little is known about this (Brand et al., 2000). In this work, we have examined possible mechanisms regulating nodule meristem development (meristems of specialized organs) in pea. Special emphasis is given to the homeodomain transcription factors of WOX family and its role in nodule organogenesis and link with CLAVATA system.

2. DEVELOPMENT OF ROOT NODULES IN PEA. LOCAL CONTROL OF NODULE ORGANOGENESIS

The formation of root nodules on legume plant roots involves two simultaneous processes: infection and nodule organogenesis. Initial events were shown to be connected with signal exchange between bacteria and host plant. In response to flavonoids exuded to the rhizosphere by legumes, the bacterial genes controlling the synthesis of nodulation factors – Nod factors are activated. Nod factors are able to trigger a cascade of events leading to the development of infection and nodule formation ((Catoira et al., 2001; Ben Amor et al., 2003; Madsen et al., 2003; Radutoiu et al., 2003; Limpens et al., 2003; Lévy et al., 2004; Mitra et al., 2004; Kaló et al., 2005; Smit et al., 2005, 2007; Heckmann et al., 2006; Kanamori et al., 2006; Middleton et al., 2007).

Most commonly the infection of plant tissue occurs through root hair cells of root epidermis. Perception of Nod factors by the host plant leads to signal cascade activation, which include membrane depolarization (Ehrhardt et al., 1992; Harris et al., 2003), calcium oscillation ("calcium waves") in perinuclear space (Ehrhardt et al., 1996; Felle et al., 1999; Engstrom et al., 2002; Charron et al., 2004), cytoskeleton structural changes (Van Brussel et al., 1992; de Ruijter et al., 1999) and, consequently, the deformation of root hairs occurs (Lerouge et al., 1990).

Responses in root epidermis cells are followed by root hairs curling such that a small bacterial microcolony becomes entrapped in the cavity of the curl. Here from the inner side of curled root hair there is local destruction of cell wall by lytic enzymes secreted by both partners. Next dividing bacteria microcolony penetrates inside the root hair cytoplasm through «invagination» of the plant cell membrane (Turgeon and Bauer 1985). As a result, infection threads (ITs) are formed with the walls similar to plant cell wall structure enclosing dividing bacteria in a polysaccharide matrix (Gage, 2004). In some legumes rhizobia are able to penetrate to root tissues through the breaks in epidermis, usually close to the site of emerging lateral root primordium (for review, see Oldroyd and Downie 2008).

Nod factor induced responses in root epidermis are coordinated with the processes occurring in root cortex - the reactivation of cell division in pericycle and cortex cells and nodule primordium formation (Oldroyd and Downie, 2008). Since Nod factors themselves do not penetrate inside root tissues and accumulate in root hair cells walls (Goedhart et al., 2000), reactivation of cell division and primordium formation seems to be regulated via interaction with endogenous regulators of plant i.e. hormones.

Pea belongs to a group of legumes forming nodules of indeterminate type, where reactivation of cell divisions occurs in the pericycle, endodermis and inner cortical cells adjacent to the xylem pole. Indeterminate nodules possess apical meristem maintaining for a long term (several months) that provides growth and renewal of nodule tissues. In contrast, in nodules of determinate type (*Lotus japonicus, Glycine max*) meristem exists only for a short period (several days). Mature determinate nodules are spherical; nodules of indeterminate type are cylindrical in shape and in some cases capable of branching. For more details concerning differences between nodules of determined and undetermined types see Ferguson et al., 2010.

A few days after rhizobia inoculation IT reaches the base of root hair, grows into root cortex then to nodule primordium cells, which continue to proliferate in parallel with IT growth into root tissues. Once IT reaches nodule primordium, the bacteria are released from the end of the IT into host cell cytoplasm through endocytosis. Bacteria are surrounded by peribacteroid membranes derived from Golgi apparatus and endoplasmic reticulum. Bacterial cells surrounded by peribacteroid membranes represent symbiosomes. Later in symbiosomes bacterial cells are differentiated into specialized symbiotic forms – bacteroids, where the synthesis of nitrogenase is activated, the enzyme catalyzing nitrogen reduction, as well as the synthesis of some enzymes required for nitrogen fixation.

3. THE ROLE OF CYTOKININ AND AUXIN IN NODULE DEVELOPMENT

The accumulating data show that change in auxin and cytokinin concentration may be secondary signal for nodule development after recognition of Nod factors (Hirsch and Fang, 1994; Mathesius et al., 1998a, 1998b, 2008; Boot et al., 1999; Pacios-Bras et al., 2003; Huo et al., 2006; Wasson et al., 2006, 2009; Grunewald et al., 2009). Experiments with an auxin-regulated reporters *GH3::GUS* and *DR5::GUS*, reflecting the auxin distribution in plants, showed the stimulation of auxin responses in dividing cells of cortex and pericycle from which nodules are initiated in legumes (Mathesius et al., 1998; Pacios-Bras et al., 2003; Huo et al., 2006; van Noorden et al., 2007). Besides that, proteomics studies showed a significant overlap (90%) in spectrum of proteins induced in response to auxin and during nodule primordium development (van Noorden et al., 2007). Suppression of genes encoding auxin efflux carriers, PIN proteins, led to a reduction in nodule number, confirming the important role of auxin in nodule development (Huo et al., 2006).

Cytokinins play a major role in the control of nodulation acting downstream of the components of Nod factor signalling (Frugier et al., 2008; Plet et al., 2011; Heckmann et al., 2011). In model legumes *Medicago truncatula* and *Lotus japonicus* components of cytokinin signaling pathway have been identified that are involved in nodule development, i.e. cytokinin receptor *Mt*CRE1/*Lj*LHK1 and B-type and A-type cytokinin response regulator RR (*Mt*RR1, *Mt*RR4, respectively) (Murray et al ., 2007; Tirichine et al., 2007). Down-regulation of cytokinin receptor gene expression *MtCRE1* using RNA-interference suppresses nodulation (Gonzalez-Rizzo et al., 2006). Besides that, gain-of-function mutant with up-regulated cytokinin receptor gene expression *LjLHK1* demonstrates spontaneous nodule formation (Tirichine et al., 2007).

It is currently proposed that upon inoculation, cytokinin signaling perturbs the expression and accumulation of auxin carriers PINs in vasculature and inner cortical cells, resulting in temporal inhibition of polar auxin transport

(PAT) to allow local auxin accumulation in the inner cortex for nodule primordium development (de Billy et al., 2001; Schnabel and Frugoli, 2004; Huo et al., 2006; Grunewald et al., 2009; Plet et al., 2011). The expression of auxin transporters seems to direct auxin transport to the developing nodules, after which normal root polar auxin transport resumes. Thus, Nod factors trigger the reaction in the root epidermis cells, leading to infection development, and at the same time, they manipulate plant endogenous regulators, such as cytokinin and auxin, controlling nodule organogenesis.

4. Systemic Control of Nodule Development. Autoregulation of Nodulation (AON) in Pea

Along with local control of nodule formation following *Rhizobium* infection there are also systemic mechanisms, regulating nodule development at the whole plant level, so-called autoregulation of nodulation (AON). Autoregulation represents the mechanism using which plant inhibits further formation of nodules on its roots after the first nodules have been formed (Caetano-Anolles and Gresshoff, 1991).

Mutants defective in AON were found in pea (*Pisum sativum* L.) and model legumes *Lotus japonicus*, *Medicago truncatula* (Carroll et al., 1985; Krusell et al., 2002; Nishimura et al., 2002; Penmetsa et al., 2003; Searle et al., 2003; Schnabel et al., 2005; 2011; Novák et al., 2011). Indeed, individual mutants were affected in a gene encoding a CLAVATA1-like leucine-rich repeat receptor-like kinase (CLV1-like LRR-RLK) (*Pssym29/ Ljhar1/ Mtsunn*) (Krusell et al., 2002; Nishimura et al., 2002; Searle et al., 2003), whereas other mutants were impaired in CLV2 LRR-RLK (*Pssym28/ Ljclv2*) (Krusell et al., 2011). All of them develop much more nodules than the corresponding wild-type plants, thus demonstrating supernodulation phenotype. Genes encoding CLV1-like receptor kinase and CLV2 involved in AON show a high percentage of sequence similarity with *CLV1* and *CLV2* genes, encoding components of CLV system in shoot apical meristem (SAM) of *Arabidopsis* and supporting the balance between cell division and differentiation (Delves et al., 1986; Olsson et al., 1989; Hamaguchi et al., 1992; Sheng and Harper, 1997; Searle et al., 2003). Both CLV1 and CLV2 in *Arabidopsis* SAM transmit the signal of the stem-cell-specific peptide hormone CLV3 to control stem cell homeostasis by forming distinct homo- or heterodimers (Oka-Kira et al., 2005; Bleckmann et al., 2010; Guo et al., 2010; Kinoshita et al., 2010; Miyazawa et al., 2010).

Grafting experiments prove that supernodulation phenotype of supernodulating mutants defected in *CLV1*-like and *CLV2* legume genes is determined by shoot, but not root, part of plant. It is assumed that plant inoculation with rhizobia activates a hypothetical root-derived signal, which comes from root to shoot, where it interacts with CLV complex, which in turn stimulates shoot-derived signal moving to roots that inhibits further nodules formation (Men et al., 2002; Nishimura et al., 2002; Oka-Kira and Kawaguchi, 2006). Recently candidates for the role of root-derived signal have been proposed: CLE-peptides have been found in the model legumes (MtCLE13, LjCLE-RS1, LjCLE-RS2, GmRIC1, GmRIC2 and GmNIC1 in *Medicago truncatula, Lotus japonicus, Glycine max*, respectively) that are specifically activated upon nodulation (Okamoto et al. 2009, Mortier et al. 2010, Reid et al. 2011). Overexpression of such nodulation-specific CLE-peptides resulted in a systemic suppression of nodulation in wild type plants, but not in supernodulaing mutants defected in CLV1-like gene (*LjHAR1, MtSUNN*) (Okamoto et al. 2009, Mortier et al. 2010). These data suggest that nodulation-specific CLE-peptides may act as ligands for CLV-like receptor kinase involved in the autoregulation of nodulation, but this hypothesis needs to be proved (Okamoto et al. 2009).

The important peculiarity of supernodulating mutants defected in CLV1-like kinase gene is nitrate-tolerance of nodulation (nitrate-tolerance symbiosis, *nts*) unlike nodulation in wild type. In wild type plants nodules are formed only under the nitrogen-poor conditions, exogenous nitrate addition inhibits nodule development due to systemic regulation (Carrol et al., 1985). Thus, the systemic mechanism associated with nitrate-inhibition of nodulation can be also mediated by CLV1-like kinase. Interestingly it was found that some CLE-peptides expressed upon nodulation are induced by nitrate addition (Okamoto et al. 2009, Reid et al. 2011). This allows speculation that *nts* phenotype of supernodulating mutants defected in CLV1-like kinase can be due to their inability to perceive nitrate-stimulated CLE-peptide. Another pea supernodulating mutant *nod4* (K301) defected in shoot-controlled AON with *nts* phenotype can be also defected in component of CLV system since it demonstrates stem fasciation, typical for other CLV mutants (Sidorova and Uzhintseva, 1995).

In addition there is a set of pea mutants defected in root-controlled AON that was demonstrated in grafting experiments. Recently one of such pea mutant *nod3* (P79 and RisFixC) has been cloned. *PsNOD3* gene appeared to be orthologous to *LjRDN1* (*ROOT-DETERMINED HYPERNODULATION*) in *L. japonicus*. *PsNOD3/LjRDN1* belongs to a small, yet uncharacterized, gene family unique to green plants that encodes a protein of unknown function expressing in vascular cylinder (Schnabel et al., 2011). Experiments have shown that it might control the production/transportation of a root-derived signal that is sent to the shoot at the onset of AON (Li et al., 2009, Novák, 2010; Schnabel et al., 2011). Along with *nod3* there are other non-allelic pea mutants (K287, K1005m) defected in root-controlled autoregulation that are still remained to be cloned (Sidorova and Shumnyi, 2003). Cloning of pea supernodulating mutants is of great interest, since it would reveal additional yet unknown components of AON system, enabling to draw the scheme of systemic regulation of nodule formation both by root and shoot.

AON is supposed to target local mechanisms of nodule development in infectes sites, limiting nodule number. It was found that *Mtsunn* mutant defected in CLV1-like gene *M. truncatula* did not show the usual nodulation-related inhibition of the shoot-to-root auxin transport (van Noorden et al., 2006). Thus, components of polar auxin transport might be targeted by AON system. However the specific molecular targets of AON and the nature of shoot-derived signals remain to be identified.

5. WOX - CLV INTERACTION IN DEVELOPMENT AND MAINTENANCE OF PLANT APICAL MERISTEMS

The components of CLV-complex in plant were initially identified using *Arabidopsis* mutants defected in SAM and flower meristem (Clark et al., 1993, 1995). First, *clv1, clv2* and *clv3* mutants were isolated characterized by common phenotype, i.e. enlarged SAM and flower meristem, increased floral organs, stem fasciations. CLV1 gene encodes Ser/Thr LRR-protein kinase, whereas CLV2 encodes LRR-protein with high similarity to CLV1 but lacking kinase domain (Clark et al., 1997; Jeong et al., 1999; Clark, 2001). The CLV1 belongs to a large class of Ser/Thr-protein kinases with leucine-rich repeats (LRR-kinase). In Arabidopsis genome 216 genes encoding LRR-kinases has been identified, dividing to 13 subfamilies (Shiu and Bleecker, 2001). The function for most of them remains unknown, various LRR-kinases being suggested to have overlaping functions. CLV3 was found to encode signal peptide of CLE-peptide family (Rojo et al., 2002). CLE-peptides (CLV3/EMBRYO-SURROUNDING REGION-related) represent recently described class of regulatory molecules. In *Arabidopsis* genome there are about 30 genes encoding putative CLE-peptides consisting of 80-120 amino acids. Conservative CLE-domain of 12 amino acids at the C-terminus of such proteins represents active form of CLE-peptide processed from the full-length protein (Oelkers et al., 2008; Wang and Fiers, 2010).

It was first believed that CLV1/CLV2 heterodimer receptor complex perceive CLE-peptide CLV3 (Brand et al., 2000; Ogawa et al., 2008) and activate signal cascade leading to restriction of meristamatic activity. Recently *crn* mutant was identified in *Arabidopsis*, exhibiting *clv*-like phenotypes (Muller et al., 2008). Corresponding *CORYNE* (*CRN*) gene encodes a membrane-associated receptor kinase without a distinct extracellular domain and it was supposed that CRN perotein can form heterodimer complex with CLV2. Using double mutants *crn clv1* it was shown that complex CRN/CLV2 might perceive CLV3 signal independently of CLV1/CLV1 homodimer complex to regulate the meristem maintenance (Muller et al., 2008).

CLV-complexes in SAM are negative regulators of meristem maintenance, restricting population of stem cells in SAM. An important function of meristem cells is to maintain its own population and to produce cells capable for differentiation, giving rise to new tissues and organs. There are two basic properties of a meristem: 1) the ability to maintain its own population of undifferentiated cells and 2) the ability to produce cells capable for differentiation and the formation of new organs. There is special zone in plant meristems, where stem cells are located – "stem cell niche", the local microenvironment, supplying the factors necessary to maintain stem cells and to suppress their differentiation (Sablowski, 2004; Singh, 2006; Scheres, 2007). It is believed that the source of such hypotetical factors within stem cell niche, suppressing the differentiation of stem cells represents a small group of cells known as "organizing center". While the daughter cells resulting from the division of stem cells are moving away organizing center out of stem cell niche, they are no longer within the action of factors produced by the organizing center, and as a consequence, undergo differentiation. This model of "stem cell niche organization" is suitable for both plant and animal stem cells (Sablowski, 2004).

In SAM stem cells are located in the central zone above their "organizing center". In the RAM organizing center is called "quiescent center" and is surrounded by stem cells, known as initial cells (Scheres, 2007). In recent years it has been shown that the key regulators of "organizing centers" in apical meristem stimulating cell proliferation and suppressing cell differentiation represent transcription factors of WOX family (WOX - from

WUSCHEL-related homeobox) (Sarkar et al., 2007, van der Graaff et al., 2009).

In *Arabidopsis* WOX family includes 15 representatives. Expression analysis of different representatives of WOX family transcription factors has shown that they regulate various stages of plant development – from the formation of embryo pattern during embryogenesis to the meristem maintenance. The general *WOX* genes function in plant development is the suppression of premature differentiation and/or stimulation of cell proliferation (van der Graaff et al., 2009).

The best studied member of the WOX family is the transcription factor WUSCHEL (WUS) expressing in the organizing center of the shoot apical meristem. WUS is essential for specification and maintenance of stem-cell proliferation in the central zone of the meristem (Mayer et al., 1998; Schoof et al., 2000, Muller et al., 2008). In *Arabidopsis* transcription factor WUS represses ARR A-type (*Arabisopsis Response Regulators A-type*) genes expression – inhibitors of cytokinin response. By this mechanism WUS stimulates cytokinin response in SAM whereby promoting cell division in SAM (Liebfried et al., 2005).

The CLV3 signaling pathway, including the CLV1/ CRN/ CLV2 receptor kinases and the CLV3 regulatory peptide, negatively regulates the *WUS* expression, thereby restricting the size of the stem cell population (Brand et al., 2000; Schoof et al., 2000). In its turn, WUS is supposed to activate CLV3 expression in SAM. Thus, CLV-WOX feedback interaction in SAM maintains the population of stem cells and thereby a balance between cell proliferation and differentiantion in SAM.

In the root apical meristem (RAM), a similar signaling system involving the WUSCHEL-related homeobox 5 (WOX5) functions in the quiescent center (QC) regulates the balance between cell division and differentiation (Kamiya et al., 2003; Haecker et al., 2004). Complementation experiments proved that WUS and WOX5 are functionally equivalent (Sarkar et al., 2007) and, therefore, could be involved in common regulatory pathways that control meristem maintenance and development in roots and shoots. In addition, a CLE peptide (CLE 40) might control *WOX5* expression domain in the RAM through the interaction with the RLK ARABIDOPSIS CRINKLY4 (ACR4) (Stahl and Simon, 2009).

6. THE COMPARISON OF NODULE DEVELOPMENT WITH OTHER DEVELOPMENTAL PROGRAMMES

According to the most common hypothesis nodule development is evolutionary related to lateral root developmental program (Mathesius et al., 2000; Bright et al., 2005). Some legumes (peanuts, some *Caesalpiniacea*) and non-legume plants involved in symbiosis with actinomycetes of genus *Frankia* develop nodules characterized by the "root" organization with a central position of vascular bundle. These nodules arise from the lateral root primordia and thus can be regarded as modified lateral roots (Wall, 2000). The similarity of lateral roots and nodules developmental programs is further confirmed by mutants affected both in lateral roots and root nodules development (Bright et al., 2005). Besides that there are abnormalities of the nodule development where ectopic root is developed from nodule meristem. Interestingly such phenotype can be caused by mutations in both plant genes *COCHLEATA* (*PsCoch*) (Ferguson and Reid, 2005), and bacteria genes (in particular, the genes controlling the biosynthesis of pyrimidines and purines) (Ferraioli et al., 2004). In addition the systemic control of nodule number and lateral root number includes common components, since supernodulating mutants also have increased lateral root number comparing with wild-type (Searle et al. 2003). All these data taken together confirm the existence of common regulatory mechanisms underlying the development of lateral roots and nodules. Apart from the fact that the development of both indeterminate nodules and lateral roots is initiated with cell divisions in pericycle cells opposite the xylem pole, another important feature of bringing together the development of nodules in lateral roots, is the presence of its own meristem. The principal differences in nodule and lateral root development are concerned with hormonal regulation of these processes. Lateral root formation is induced by auxin that stimulates pericycle cell proliferation leading to lateral root primordium development. It was shown that *WOX5* gene induction upon first pericycle cell divisions is activated by auxin (Gonzali et al., 2005; Imin et al., 2007; Chen et al., 2009). In contrast to auxin, cytokinin suppresses lateral root development. Local increase in cytokinin concentration in pericycle cells (e.g., caused by overexpression of the cytokinin gene biosynthesis izopentiniltransferazy) blocks lateral root primordia formation. Exogenous cytokinin addition causes changes in orientation of cell divisions in emerging lateral root primordium, altering its structure, which results in a "flat primordium" of lateral root (Laplaze et al., 2007). In contrast to lateral roots, it was shown that cytokinin stimulates cell divisions during nodule primordium development in root cortex.

CLE-peptides play a special role in the regulation of cell proliferation in plants. In addition to their role in the regulation of "regular" meristem (SAM, RAM) development (in particular, CLV3, CLE40), these regulators also control cell proliferation during the development of symbiotic root nodules in legumes. What are the factors interacting with CLE-peptides outside the "regular" meristems?

WOX5 homeodomain transcription factor is proposed to be the important regulator of root meristem maintenance in *Arabidopsis*, being expressed in quiesence center («organizer») of root meristem (Haecker et al., 2004). Since the developmental programs of lateral roots and nodules are evolutionarily related to each other, the WOX5 may be also involved in the control of nodule organogenesis. How could the components of CLV system (CLE-peptides, CLV receptor kinase) influence on WOX5 expression upon nodule development? Answer to this question can help to find out the mechanisms regulating different types of meristems. Pea symbiotic nodule development may be considered as a suitable model to answer specified question.

7. COMMON ASPECTS IN NODULE DEVELOPMENT AND OTHER TYPES OF PLANT MERISTEM DEVELOPMENT

Evidences that CLV1-like and CLV2 receptors are implicated in AON regulation of nodulation suggest the existence of common regulatory pathways controlling apical and nodule meristem development. Another argument in favor of this hypothesis is the finding that nodule-specific CLE peptides is involved in AON regulation. As mentioned above the target of CLV complex in SAM might be homeobox transcription factor of WOX family. Recently, analysis of the temporal *WOX5* expression in *Pisum sativum* L. during nodulation with quantitative reverse-transcription PCR and promoter-reporter fusion revealed that the *WOX5* gene is expressed during nodule organogenesis, suggesting that *WOX* genes are common regulators of cell proliferation in different systems (Osipova et al., in press). *Agrobacterium rhizogenes* - mediated transformation of pea by hypocotyls stabbing method allowed obtaining transgenic roots expressing *pMtWOX5:GUS* construct. Based on the histological analysis of such plants the *WOX5* expression was shown to be increased early after *Rhizobium* inoculation and this gene is highly expressed in proliferating cells of nodule primordia (Osipova et al., in press). At later stages of nodule development *WOX5* expression gradually decreases and is restricted to the apical region of fully grown nodule.

To study the possible interaction of WOX5 with CLAVATA-like system upon nodulation *WOX5* expression has been analysed in pea supernodulating mutants defected in shoot-determined AON: *Pssym29* defected in *PsCLV1*-like gene, *Pssym28* defected in *PsCLV2* gene. Analysis of pea mutants defective in AON revealed that the signaling via shoot-acting CVL1 and CVL2 proteins represses the *WOX5* expression in developing nodules (Osipova et al., in press). Hence, a conserved WUS/WOX-CLV regulatory system might control cell proliferation and differentiation not only in the root and shoot apical meristems, but also in nodule meristems.

CONCLUSION

So the analysis of *WOX5* expression upon nodulation revealed the existence of common mechanisms controlling cell proliferation by *WOX* genes in apical meristems and during the development of specialized organs – legume symbiotic nodules (Table). Upon nodule development *WOX5* gene expression is also regulated by CLAVATA-like system, similar to its paralogue, *WUS* active in SAM. However, the peculiarity of *WOX5* expression regulation upon nodulation is that it is regulated by CLV-like system by a long distance pathway - and the molecular mechanisms of such long-distance pathway are still to be elucidated.

Table. Common group of factors, regulating different types of meristems

Meristem type	WOX transcription factor	CLE-peptide	Receptor Set/Thr kinase
SAM	WUS	CLV3	CLV1/CLV1, CRN/CLV2
RAM	WOX5	CLE40, CLE19	ACR4 CRN/CLV2 -?
Nodule meristem	WOX5	Nodulation-specific CLE-peptides (MtCLE13, LjCLE-RS1, LjCLE-RS2)	CLV1-like kinase (PsSYM29) CLV2 (PsSYM28)

Besides interaction with components of CLV complex member of WOX family are interconnected with hormones action in meristems. So *WOX5* gene expression is activated by auxin in RAM (Gonzali et al., 2005; Imin et al., 2007; Chen et al., 2009) whereas WUS stimulates cytokinin signalling in SAM (Leibfries et al, 2005). In general these findings suggest that WOX transcription factors could be converging point for both local hormonal regulation of meristem function and systemic regulation, integrating information from different signals to regulate meristem activity.

REFERENCES

Ben Amor, B, Shaw, S.L., Oldroyd, G.E.D., Maillet, F,, Penmetsa, R.V., Cook, D., Long, S.R., Dénarié, J., Gough, C. (2003) The NFP locus of *Medicago truncatula* controls an early step of Nod factor signal transduction upstream of a rapid calcium flux and root hair deformation. *Plant J* 34, 495-506.

Bleckmann, A., Weidtkamp-Peters, S., Seidel, C.A.M, Simon, R. (2010) Stem cell signaling in *Arabidopsis* requires CRN to localize CLV2 to the plasma membrane. *Plant Physiol* 152,166-176.

Boot, K.J.M, van Brussel, A.A.N., Tak, T., Spaink, H.P., Kijne, J.W. (1999) Lipochitin oligosaccharides from Rhizobium leguminosarum bv. viciae reduce auxin transport capacity in *Vicia sativa* subsp. nigra roots. *Mol Plant-Microbe Interact* 12, 839-844.

Borisov, A.Y., Barmicheva, E.M., Jacobi, L.M., Tsyganov, V.E., Voroshilova, V.A., Tikhonovich, I.A. Pea (*Pisum sativum* L.) mendelian genes controlling development of nitrogen-fixing nodules and arbuscular mycorrhiza (2000) *Czech Journal of Genetics and Plant Breeding,* 36, 106–110.

Brand, U., Fletcher, J.C., Hobe, M., Meyerowitz, E.M., Simon, R. (2000) Dependence of stem cell fate in *Arabidopsis* on a feedback loop regulated by CLV3 activity. *Science,* 289, 617-619.

Bright, L.J., Liang, Y., Mitchell, D.M., Harris, J.M. (2005) The LATD gene of Medicago truncatula is required for both nodule and root development Mol. *Plant-Microbe Interact.* 18, 521-532.

Caetano-Anollés, G., Gresshoff, P.M. (1991) Plant genetic control of nodulation. *Annu Rev Microbiol* 45. 345-382.

Carroll, B.J., McNeil, D.L., Gresshoff, P.M. (1985) Isolation and properties of soybean [Glycine max (L.) Merr.] mutants that nodulate in the presence of high nitrate concentrations. *Proc Natl Acad Sci USA* 82, 4162-4166.

Catoira, R., Timmers, A.C.J., Maillet, F., Galera, C., Penmetsa, R.V., Cook, D., Dénarié, J, Gough, C. (2001) The HCL gene of *Medicago truncatula* controls *Rhizobium*-induced root hair curling. *Development* 128, 1507-1518.

Chen, S.-K., Kurdyukov, S., Kereszt, A., Wang, X.-D., Gresshoff, P.M., Rose, R.J. (2009) The association of homeobox gene expression with stem cell formation and morphogenesis in cultured Medicago truncatula. *Planta* 230. 827-840.

Clark, S.E. (2001) Cell signaling at the shoot meristem. *Nat Rev Mol Cell Biol,* 2, 276-284.

Clark, S.E., Running, M.P. and Meyerowitz, E.M. (1993) CLAVATA1, a regulator of meristem and flower development in Arabidopsis. *Development,* 119, 397–418.

Clark, S.E., Running, M.P. and Meyerowitz, E.M. (1995) CLAVATA3 is a specific regulator of shoot and floral meristem development affecting the same processes as CLAVATA1. *Development,* 121, 2057–2067. .

Clark, S.E., Williams, R.W., Meyerowitz, E.M. (1997) The CLAVATA1 gene encodes a putative receptor kinase that controls shoot and floral meristem size in Arabidopsis. *Cell* 89, 575-585.

Delves, A.C., Mathews, A., Day, D.A., Carter, A.S., Carroll, B.J., Gresshoff, P.M. (1986) Regulation of the soybean-*Rhizobium* nodule symbiosis by shoot and root factors. *Plant Physiol* 82, 588-590.

de Ruijter, N.C.A., Bisseling, T., Emons, A.M.C. (1999) *Rhizobium* Nod factors induce an increase in sub-apical fine bundles of actin filaments in *Vicia sativa* root hairs within minutes. *Mol. Plant Microbe Interact.* 12,829-832.

Ehrhardt, D.W., Atkinson, E.M., Long, S.R. (1992) Depolarization of alfalfa root hair membrane potential by *Rhizobium meliloti* Nod factors. *Science.* 256. 998–1000.

Ehrhardt, D.W., Wais, R., Long, S.R. (1996) Calcium spiking in plant root hairs responding to Rhizobium nodulation signals. *Cell.* 85, 673–681.

Engstrom, E.M., Ehrhardt, D.W., Mitra, R.M., Long, S.R. (2002) Pharmacological analysis of Nod factor-induced calcium spiking in *Medicago truncatula*: evidence for the requirement of type IIA calcium pumps and phosphoinositide signaling. *Plant Physiol.* 128, 1390-1401.

Felle, H.H., Kondorosi, E., Kondorosi, Á., Schultze, M. (1999) Nod factors modulate the concentration of cytosolic free calcium differently in growing and non-growing root hairs of *Medicago sativa* L. *Planta.* 209, 207-212.

Ferguson, B.J., Indrasumunar, A., Hayashi, S., Lin, M.-H., Lin,.Y.-H., Reid, D.E., Gresshoff, P.M. (2010) Molecular analysis of legume nodule development and autoregulation. *J Integr Plant Biol* 52, 61-76.

Ferguson, B.J., Reid, J.B. (2005) Cochleata: Getting to the Root of Legume Nodules. *Plant Cell Physiol* 46(9), 1583–1589.

Ferraioli, S., Tatè, R., Rogato, A., Chiurazzi, M., Patriarca, E.J. (2004) Development of ectopic roots from abortive nodule primordia. *Mol Plant Microbe Interact.* 10. 1043-50.

Frugier, F., Kosuta, S., Murray, J.D., Crespi, M., Szczyglowski, K. (2008) Cytokinin: secret agent of symbiosis. *Trends Plant Sci* 13, 115-120.

Gage, D.J. (2004) Infection and invasion of roots by symbiotic, nitrogen-fixing rhizobia during nodulation of temperate legumes. *Microbiol Mol Biol Rev.* 68, 280-300.

Goedhart, J., Hink, M.A., Visser, A.J., Bisseling, T., Gadella, T.W. Jr. (2000) *In vivo* fluorescence correlation microscopy (FCM) reveals accumulation and immobilization of Nod factors in root hair cell walls. *Plant J.* 21(1), 109-19.

Gonzalez-Rizzo, S., Crespi, M., Frugier, F. (2006) The Medicago truncatula CRE1 cytokinin receptor regulates lateral root development and early symbiotic interaction with Sinorhizobium meliloti. *Plant Cell* 18, 2680-2693.

Gonzali, S., Novi, G., Loreti, E., Paolicchi, F., Poggi, A., Alpi, A., Perata, P. (2005) A turanose-insensitive mutant suggests a role for WOX5 in auxin homeostasis in Arabidopsis thaliana. *Plant J.* 44, 633-645.

Grunewald, W., van Noorden, G., Van Isterdael, G., Beeckman, T., Gheysen, G., Mathesius, U. (2009) Manipulation of auxin transport in plant roots during *Rhizobium* symbiosis and nematode parasitism. *Plant Cell* 21, 2553-2562.

Haecker, A., Groß-Hardt, R., Geiges, B., Sarkar, A., Breuninger, H., Herrmann, M., Laux, T. (2004) Expression dynamics of *WOX* genes mark cell fate decisions during early embryonic patterning in Arabidopsis thaliana. *Development* 131, 657-668.

Hamaguchi, H., Kokubun, M., Akao, S. (1992) Shoot control of nodulation is modified by the root in the supernodulating soybean mutant En6500 and its wild-type parent cultivar Enrei. *Soil Sci Plant Nutr* 38, 771-774.

Harris, J.M., Wais, R., Long, S.R. (2003) *Rhizobium*-induced calcium spiking in Lotus japonicas. *Mol. Plant Microbe Interact.* 16, 335-341.

Heckmann, A.B., Lombardo, F., Miwa, H., Perry, J.A., Bunnewell, S., Parniske, M., Wang, T.L., Downie, J.A. (2006) Lotus japonicus nodulation requires two GRAS domain regulators, one of which is functionally conserved in a non-legume. *Plant Physiol* 142, 1739-1750.

Heckmann, A.B., Sandal, N., Bek, A.S., Madsen, L.H., Jurkiewicz, A., Nielsen, M.W., Tirichine, L., Stougaard, J. (2011) Cytokinin induction of root nodule primordia in Lotus japonicus is regulated by a mechanism operating in the root cortex. *Mol Plant-Microbe Interact in press* (DOI: 10.1094/MPMI-05-11-0142).

Hirsch, A.M. and Fang, Y. (1994) Plant hormones and nodulation: what's the connection? *Plant Mol Biol* 26, 5-9.

Huo, X., Schnabel, E., Hughes, K., Frugoli, J. (2006) RNAi phenotypes and the localization of a protein::GUS fusion imply a role for Medicago truncatula PIN genes in nodulation. *J Plant Growth Regul* 25, 156-165.

Imin, N., Nizamidin, M., Wu, T., Rolfe, B.G. (2007) Factors involved in root formation in Medicago truncatula. *J Experimental Botany.* 58, 439-451.

Jeong, S., Trotochaud, A.E. and Clark, S.E. (1999) The Arabidopsis CLAVATA2 gene encodes a receptor-like protein required for the stability of the CLAVATA1 receptor-like kinase. *Plant Cell,* 11, 1925–1934.

Kaló, P., Gleason, C., Edwards, A., Marsh, J., Mitra, R.M., Hirsch, S., Jakab, J., Sims, S., Long, S.R., Rogers, J., Kiss, G.B., Downie, J.A., Oldroyd, G.E.D. (2005) Nodulation signaling in legumes requires NSP2, a member of the GRAS family of transcriptional regulators. *Science* 308, 1786-1789.

Kamiya, N., Nagasaki, H., Morikami, A., Sato, Y., Matsuoka, M. (2003) Isolation and characterization of a rice WUSCHEL-type homeobox gene that is specifically expressed in the central cells of a quiescent center in the root apical meristem. *Plant J* 35, 429-441.

Kanamori, N., Madsen, L.H., Radutoiu, S., Frantescu, M., Quistgaard, E.M.H., Miwa, H., Downie, J.A., James, E.K., Felle, H.H., Haaning, L.L., Jensen, T.H., Sato, S., Nakamura, Y., Tabata, S., Sandal, N., Stougaard, J. (2006) A nucleoporin is required for induction of Ca2+ spiking in legume nodule development and essential for rhizobial and fungal symbiosis. *Proc Natl Acad Sci USA* 103, 359-364.

Kosslak, R.M., Bohlool, B.B. (1984) Suppression of nodule development of one side of a split-root system of soybeans caused by prior inoculation of the other side. *Plant Physiol* 75, 125-130.

Krusell, L., Madsen, L.H., Sato, S., Aubert, G., Genua, A., Szczyglowski, K., Duc, G., Kaneko, T., Tabata, S., de Bruijn, F., Pajuelo, E., Sandal, N., Stougaard, J. (2002) Shoot control of root development and nodulation is mediated by a receptor-like kinase. *Nature* 420, 422-426.

Krusell, L., Sato, N., Fukuhara, I., Koch, B.E.V., Grossmann, C., Okamoto, S., Oka-Kira, E., Otsubo, Y., Aubert, G., Nakagawa, T., Sato, S., Tabata, S., Duc, G., Parniske, M., Wang, T.L., Kawaguchi, M., Stougaard, J. (2011) The Clavata2 genes of pea and Lotus japonicus affect autoregulation of nodulation. *Plant J* 65, 861-871.

Laplaze, L., Benkova, E., Casimiro, I., Maes, L., Vanneste, S., Swarup, R., Weijers, D., Calvo, V., Parizot, B., Herrera-Rodriguez, M.B., Offringa, R., Graham, N., Doumas, P., Friml, J., Bogusz, D., Beeckman, T., Bennett, M. (2007) Cytokinins act directly on lateral root founder cells to inhibit root initiation. *Plant Cell.*19, 3889–3900.

Lévy, J., Bres, C., Geurts, R., Chalhoub, B., Kulikova, O., Duc, G., Journet, E.-P., Ané, J.-M., Lauber, E., Bisseling, T., Dénarié, J., Rosenberg, C., Debellé, F. (2004) A putative Ca2+ and calmodulin-dependent protein kinase required for bacterial and fungal symbioses. *Science* 303, 1361-1364.

Lerouge, P., Roche, P., Faucher, C., Maillet, F., Truchet, G., Promé, J.C., Dénarié, J. (1990) Symbiotic host-specificity of Rhizobium meliloti is determined by a sulphated and acylated glucosamine oligosaccharide signal. *Nature.* 19. 344, 781-4.

Li, D., Kinkema, M., Gresshoff, P.M. (2009) Autoregulation of nodulation (AON) in *Pisum sativum* (pea) involves signalling events associated with both nodule primordia development and nitrogen fixation. *J Plant Physiol* 166, 955-967.

Leibfried, A., To, J.P., Busch, W., Stehling, S., Kehle, A., Demar, M., Kieber, J.J., Lohmann, J.U. (2005) WUSCHEL controls meristem function by direct regulation of cytokinin-inducible response regulator. *Nature* 438, 1172-5.

Limpens, E., Franken, C., Smit, P., Willemse, J., Bisseling, T., Geurts, R. (2003) LysM domain receptor kinases regulating rhizobial Nod factor-induced infection. *Science* 302, 630-633.

Madsen, E.B., Madsen, L.H., Radutoiu, S., Olbryt, M., Rakwalska, M., Szczyglowski, K., Sato, S., Kaneko, T., Tabata, S., Sandal, N., Stougaard J. (2003) A receptor kinase gene of the LysM type is involved in legume perception in rhizobial signals. *Nature* 425, 637-640.

Mathesius, U., Schlaman, H.R.M., Spaink, H.P., Sautter, C., Rolfe, B.G., Djordjevic, M.A. (1998) Auxin transport inhibition precedes root nodule formation in white clover roots and is regulated by flavonoids and derivatives of chitin oligosaccharides. *Plant J* 14, 23-34.

Mathesius, U., Weinman, J.J., Rolfe, B.G., Djordjevic, M.A. (2000) Rhizobia can induce nodules in white clover by "hijacking" mature cortical cells activated during lateral root development. *Mol Plant-Microbe Interact* 13, 170-182.

Mayer, K.F.X., Schoof, H., Haecker, A., Lenhard, M., Jürgens, G., Laux, T. (1998) Role of WUSCHEL in regulating stem cell fate in the Arabidopsis shoot meristem. *Cell* 95, 805-815.

Middleton, P.H., Jakab, J., Penmetsa, R.V., Starker, C.G., Doll, J., Kaló, P., Prabhu, R., Marsh, J.F., Mitra, R.M., Kereszt, A., Dudas, B., VandenBosch, K., Long, S.R., Cook, D.R., Kiss, G.B., Oldroyd, G.E.D. (2007) An ERF transcription factor in Medicago truncatula that is essential for Nod factor signal transduction. *Plant Cell* 19, 1221-1234.

Mitra, R.M., Gleason, C.A., Edwards, A., Hadfield, J., Downie, J.A., Oldroyd, G.E.D., Long, S.R. (2004) A Ca2+/calmodulin-dependent protein kinase required for symbiotic nodule development, gene identification by transcript-based cloning. *Proc Natl Acad Sci USA* 101, 4701-4705

Miyazawa, H., Oka-Kira, E., Sato, N., Takahashi, H., Wu, G.-J., Sato, S., Hayashi, M., Betsuyaku, S., Nakazono, M., Tabata, S., Harada, K., Sawa, S., Fukuda, H., Kawaguchi, M. (2010) The receptor-like kinase KLAVIER mediates systemic regulation of nodulation and non-symbiotic shoot development in Lotus japonicus. *Development* 137, 4317-4325.

Mortier, V., Den Herder, G., Whitford, R., Van de Velde, W., Rombauts, S., D'haeseleer, K., Holsters, M., Goormachtig, S. (2010) CLE peptides control Medicago truncatula nodulation locally and systemically. *Plant Physiol* 153, 222-237.

Müller, R., Bleckmann, A., Simon, R. (2008) The receptor kinase CORYNE of Arabidopsis transmits the stem cell-limiting signal CLAVATA3 independently of CLAVATA1. *Plant Cell* 20, 934-946.

Murray, J.D., Karas, B.J., Sato, S., Tabata, S., Amyot, L., Szczyglowski, K. (2007) A cytokinin perception mutant colonized by Rhizobium in the absence of nodule organogenesis. *Science.* 315, 101-104.

Nishimura, R., Hayashi, M., Wu, G.-J., Kouchi, H., Imaizumi-Anraku, H., Murakami, Y., Kawasaki, S., Akao, S., Ohmori, M., Nagasawa, M., Harada, K., Kawaguchi, M. (2002) HAR1 mediates systemic regulation of symbiotic organ development. *Nature* 420, 426-429.

Novák, K. (2010) Early action of pea symbiotic gene NOD3 is confirmed by adventitious root phenotype. *Plant Sci* 179, 472-478.

Novák, K., Lisá, L., Škrdleta, V. (2011) Pleiotropy of pea RisfixC supernodulation mutation is symbiosis-independent. *Plant Soil* 342, 173-182.

Ogawa, M., Shinohara, H., Sakagami, Y., Matsubayashi, Y. (2008) Arabidopsis CLV3 peptide directly binds CLV1 ectodomain. *Science* 319, 294-294 [Erratum Science 319, 901].

Oka-Kira, E., Tateno, K., Miura, K., Haga, T., Hayashi, M., Harada, K., Sato, S., Tabata, S., Shikazono, N., Tanaka, A., Watanabe, Y., Fukuhara, I., Nagata, T., Kawaguchi, M. (2005) klavier (klv), a novel hypernodulation mutant of Lotus japonicus affected in vascular tissue organization and floral induction. *Plant J* 44, 505-515.

Oka-Kira, E., Kawaguchi, M. (2006) Long-distance signaling to control root nodule number. *Curr Opin Plant Biol* 9, 496-502.

Okamoto, S., Ohnishi, E., Sato, S., Takahashi, H., Nakazono, M., Tabata, S., Kawaguchi, M. (2009) Nod factor/nitrate-induced CLE genes that drive HAR1-mediated systemic regulation of nodulation. *Plant Cell Physiol* 50, 67-77.

Oldroyd, G.E.D., Downie, J.A (2008) Coordinating nodule morphogenesis with rhizobial infection in legumes. *Annu Rev Plant Biol* 59, 519-546.

Olsson, J.E., Nakao, P., Bohlool, B.B., Gresshoff, P.M. (1989) Lack of systemic suppression of nodulation in split root systems of supernodulating soybean (Glycine max [L.] Merr.) mutants. *Plant Physiol* 90, 1347-1352.

Osipova, M. A., Mortier, V., Tsyganov, V. E., Tikhonovich, I. A., Lutova, L.A., Dolgikh, E. A., Goormachtig, S. *WUSCHEL-RELATED HOMEOBOX5* gene expression and interaction of CLE peptides with components of the systemic control add two pieces to the puzzle of autoregulation of nodulation. *Plant Physiol.* In press.

Pacios-Bras, C., Schlaman, H.R.M., Boot, K., Admiraal, P., Langerak, J.M., Stougaard, J., Spaink, H.P. (2003) Auxin distribution in Lotus japonicus during root nodule development. *Plant Mol Biol* 52, 1169-1180.

Penmetsa, R.V., Frugoli, J.A., Smith, L.S., Long, S.R., Cook, D.R. (2003) Dual genetic pathways controlling nodule number in Medicago truncatula. *Plant Physiol* 131, 998-1008.

Plet, J., Wasson, A., Ariel, F., Le Signor, C., Baker, D., Mathesius, U., Crespi, M., Frugier, F. (2011) MtCRE1-dependent cytokinin signaling integrates bacterial and plant cues to coordinate symbiotic nodule organogenesis in Medicago truncatula. *Plant J* 65, 622-633.

Radutoiu, S., Madsen, L.H., Madsen, E.B., Felle, H.H., Umehara, Y., Grønlund, M., Sato, S., Nakamura, Y., Tabata, S., Sandal, N., Stougaard, J. (2003) Plant recognition of symbiotic bacteria requires two LysM receptor-like kinases. *Nature* 425, 585-592.

Reid, D.E., Ferguson, B.J., Gresshoff, P.M. (2011) Inoculation- and nitrate-induced CLE peptides of soybean control NARK-dependent nodule formation. *Mol Plant-Microbe Interact* 24, 606-618.

Rojo, E., Sharma, V.K., Kovaleva, V., Raikhel, N.V. and Fletcher, J.C. (2002) CLV3 is localized to the extracellular space, where it activates the Arabidopsis CLAVATA stem cell signaling pathway. *Plant Cell,* 14, 969–977.

Sablowski, R. (2004) Plant and animal stem cells, conceptually similar, molecularly distinct? *Trends Cell Biol.* 14, 605-611.

Sarkar, A.K., Luijten, M., Miyashima, S., Lenhard, M., Hashimoto, T., Nakajima, K., Scheres, B., Heidstra, R., Laux, T. (2007) Conserved factors regulate signalling in Arabidopsis thaliana shoot and root stem cell organizers. *Nature* 446, 811-814.

Scheres, B. (2007) Stem-cell niches, nursery rhymes across kingdoms. *Nat Rev Mol Cell Biol.* 8, 345-54.

Schnabel, E., Journet, E.-P., de Carvalho-Niebel, F., Duc, G., Frugoli, J. (2005) The Medicago truncatula SUNN gene encodes a CLV1-like leucine-rich repeat receptor kinase that regulates nodule number and root length. *Plant Mol Biol* 58, 809-822.

Schnabel, E.L., Frugoli, J. (2004) The PIN and LAX families of auxin transport genes in Medicago truncatula. *Mol Genet Genomics* 272, 420-432.

Schnabel, E.L., Kassaw, T.K., Smith, L.S., Marsh, J.F., Oldroyd, G.E., Long, S.R., Frugoli, J.A. (2011) The ROOT DETERMINED NODULATION1 gene regulates nodule number in roots of Medicago truncatula and defines a highly conserved, uncharacterized plant gene family. *Plant Physiol* 157, 328-340.

Schoof, H., Lenhard, M., Haecker, A., Mayer K.F.X., Jürgens, G., Laux, T. (2000) The stem cell population of Arabidopsis shoot meristem is maintained by a regulatory loop between the CLAVATA and WUSCHEL genes. *Cell* 100, 635-644.

Searle, I.R., Men, A.E., Laniya, T.S., Buzas, D.M., Iturbe-Ormaetxe, I., Carroll, B.J., Gresshoff, P.M. (2003) Long-distance signaling in nodulation directed by a CLAVATA1-like receptor kinase. *Science* 299, 109-112.

Sheng, C., Harper, J.E. (1997) Shoot versus root signal involvement in nodulation and vegetative growth in wild-type and hypernodulating soybean genotypes. *Plant Physiol* 113, 825-831.

Shiu, S.-H. and Bleecker, A.B. (2001) Receptor-like kinases from Arabidopsis form a monophyletic gene family related to animal receptor kinases. *Proc Natl Acad Sci U S A.* 98(19), 10763–10768.

Smit, P., Limpens, E., Geurts, R., Fedorova, E., Dolgikh, E., Gough, C., Bisseling, T. (2007) Medicago LYK3, an entry receptor in rhizobial nodulation factor signaling. *Plant Physiol* 145, 183-191.

Stahl, Y., Simon, R. (2009) Is the Arabidopsis root niche protected by sequestration of the CLE40 signal by its putative receptor ACR4? *Plant Signal Behav* 4, 634-635.

Sidorova, K.K. and Shumnyi, V.K. (2003) A Collection of Symbiotic Mutants in Pea Pisum sativum L., Creation and Genetic Study. *Russian Journal of Genetics.* 39,406-413.

Sidorova, K.K. and Uzhintseva, L.P. (1995). Mapping of nod-4, a new hypernodulating mutant in pea. *Pisum Genetics.* 27, 21.

Tirichine, L., Sandal, N., Madsen, L.H., Radutoiu, S., Albrektsen, A.S., Sato, S., Asamizu, E., Tabata, S., Stougaard, J. (2007) A gain-of-function mutation in a cytokinin receptor triggers spontaneous root nodule organogenesis. *Science* 315, 104-107.

Turgeon, B.G. and Bauer, W. D. (1985) Ultrastructure of infection-thread development during the infection of soybean by Rhizobium japonicum. *Planta. 163*, 328-349.

Tsyganov, V.E., Voroshilova, V.A., Priefer, U.B., Borisov, A.Y., Tikhonovich, I.A. (2002). Genetic dissection of the initiation of the infection process and nodule tissue development in the Rhizobium–pea (Pisum sativum L.) symbiosis. *Annals of Botany* 89, 357-366.

van Brussel, A.A.N., Planque, K., Quispel, A. (1977) The wall of *Rhizobium leguminosarum* in bacteroid and free-living forms. *Journal of General Microbiology.* 101, 51-56.

van der Graaff, E., Laux, T., Rensing, S.A. (2009) The WUS homeobox-containing (WOX) protein family. *Genome Biol* 10, 248.

van Noorden, G.E., Ross, J.J., Reid, J.B., Rolfe, B.G., Mathesius, U. (2006) Defective long-distance auxin transport regulation in the *Medicago truncatula super numeric nodules* mutant. *Plant Physiol* 140, 1494-1506.

Wang, G., Fiers, M. (2010) CLE peptide signaling during plant development. *Protoplasma* 240: 33-43.

In: Peas   ISBN: 978-1-61942-866-9
Editors: A. Comstock and B. Lothrop   © 2012 Nova Science Publishers, Inc.

Chapter 4

DEVELOPING FALL-SOWN PEA CULTIVARS AS AN ANSWER TO THE CHALLENGES OF CLIMATIC CHANGES

Aleksandar Mikić[1], Vojislav Mihailović[1],
Branko Ćupina[2], Isabelle Lejeune-Henaut[3],
Eric Hanocq[3], Gérard Duc[4], Kevin McPhee[5],
Frederick L. Stoddard[6], Valentin Kosev[7], Đorđe Krstić[2],
Svetlana Antanasović[2] and Živko Jovanović[8]

[1]Institute of Field and Vegetable Crops, Novi Sad, Serbia
[2]University of Novi Sad, Faculty of Agriculture, Novi Sad, Serbia
[3]Institut National de la Recherche Agronomique, Unité Mixte
de Recherche en Génétique et Ecophysiologie des Légumineuses
à Graines, Dijon, France
[4]Institut National de la Recherche Agronomique, Unité Mixte
de Recherche Stress Abiotiques et Différenciation des
Végétaux Cultivés, Estrées-Mons, France
[5]North Dakota State University, Fargo, US
[6]University of Helsinki, Department of Agricultural Sciences,
Helsinki, Finland
[7]Institute for Forage Crops, Pleven, Bulgaria
[8]University of Belgrade, Institute for Molecular Genetics
and Genetic Engineering, Belgrade, Serbia

ABSTRACT

Pea is considered rather well adapted to wide temperature ranges, with seedlings able to survive even -20 °C. From a physiological viewpoint, pea becomes tolerant to frost if first exposed to low non-freezing temperatures, causing the so-called cold acclimation. Delayed floral initiation helps some forage pea genotypes to escape the main winter freezing periods, as susceptibility to frost increases during the transition to the reproductive state. The oldest winter pea cultivars carry the dominant allele, *Hr*, although some bear *hr*. They are generally characterized by prominent winter hardiness and a long growing season, from sowing in early October until either cutting for forage production in late May or harvesting seeds in mid-July. The average forage yields in the winter forage pea cultivars often exceed 45 t ha^{-1} of green forage, 9 t ha^{-1} of forage dry matter and 1700 kg ha^{-1} of forage crude protein. Modern dry pea cultivars have advanced winter hardiness and enhanced dry grain production. They are already in use in other temperate regions in both Europe, especially France, and the USA. One of the strategic advantages of fall-sown dry pea cultivars of recent release is their significantly improved earliness. These cultivars are regularly at least one week earlier than winter barley, providing many farmers with the novel opportunity of not having to choose between pea and cereals, since many have only one combine harvester available and give priority to their cereals. Furthermore, fall-sown dry pea cultivars may have increased grain dry matter crude protein content and it is possible to merge winter hardiness and low content of anti-nutritional factors. Low thousand seed weight, not exceeding 200 g, and a population density of 75-80 plants m^{-2} provide inexpensive sowing. All these outcomes should result in an increased area and production of dry pea in many temperate regions. In the end, growing winter-hardy pea cultivars also mean a significant shift into the wetter half of the year and thus mitigating more and more prominent and unpredictable effects of spring droughts, demonstrating an efficient answer to the challenges of climatic changes.

Keywords: abiotic stress, climatic changes, drought, low temperatures, pea, *Pisum sativum*, sowing time, winter hardiness.

INTRODUCTION

Pea (Pisum sativum L.) is one of the most important annual cool season legumes in the world today (Mihailović and Mikić 2010). Producing pea (*Pisum sativum* L.) is one of the least expensive and at the same time most quality answers to a perennial demand for plant protein by animal husbandry (Maxted and Ambrose 2000). Pea also represents a valuable addition to or a complete replacement for soybean meal in the years with less favorable conditions for the cultivation of the latter (Mikić et al. 2003). From an agronomic point of view, dry pea denotes the cultivars that are used exclusively for grain, with white flowers, more or less round grains, yellow-white or blue-green testa and a very low or low content of anti-nutritional factors, especially trypsin inhibitors (Mikić et al. 2009). The exceptions are the dry pea cultivars of previous generations, as well as forage pea cultivars, with purple flowers and olive or marble testa (Carrouée 1993; Mihailović et al. 2004).

The total harvested area under the pea crop in the world is more than 6,000,000 ha, with the grain production of more than 12,000,000 t (FAOSTAT 2007). While countries such as Canada record a constant trend of increasing and maintaining a high level of dry pea production, the harvested area under the pea crop in most of the European countries is faced with a serious decrease. The loss of interest in the pea cultivation by European farmers is mainly due to the outcomes of the previous aspects of the Common Agricultural Policy that favors cereals and brassicas in comparison to grain legumes. Luckily enough, following the most recent decisions by the EU government, it is to be expected that the production of pea and other grain legumes in Europe should witness a new revival (Burstin 2009). Their total area in Europe may and should be increased due to many environmental, economic and social reasons. Recent agroeconomic research in contrasting regions of the European Union confirmed that pea and other grain legumes may profitably be included in diverse crop rotations every 3–6 years (Nemecek et al. 2008). Achieving this strategic goal requests the vivid and strengthened interactions between genetics, breeding, agroecology and agronomy.

One of the most prominent features of the climatic changes affecting temperate regions is a rather irregularly distributed and generally decreased amount of rainfall. This seriously endangers the cultivation of spring-sown crops, even the earliest one such as cereals or pea, since the drought may appear in the stages of its flowering or grain development, significantly reducing yields. Among the primarily recognized strategies for mitigating the disastrous effects of drought on grain legumes such as pea or faba bean (*Vicia faba* L.) is the development of fall-sowing cultivars with good winter hardiness and prominent earliness, that simultaneously contributes to increasing the cultivation area under this crop and better utilization of natural resources (Duc 1997, Mikić et al. 2011a).

GENETIC BACKGROUND OF WINTER HARDINESS IN PEA

Pea is regarded as rather well adapted to wide ranges of temperature. Its seedlings are able to survive up to -20 °C (Shereena and Salim 2006). Germination and emergence in pea are not generally associated with seed or flower color. However, pink-flowered genotypes demonstrate better germination and emergence at 5°C than those with white flowers (Sincik et al. 2004) and thus explain why wild populations of the pea subspecies *arvense* are often prominently rather winter hardy. Purely physiologically, pea becomes tolerant to frost if first exposed to low but non-freezing temperatures, leading to the so-called cold acclimation. In such cases, regular light intensity improves the freezing tolerance. Also, there is a close relationship between the soluble sugar concentration of leaves in the eve of the frost and the degree of freezing tolerance (Bourion et al. 2003). At the same time, pea plants have ability to modulate their photosynthetic rate during growth at low temperatures and adjust it to the extent needed for survival (Yordanov et al. 1996).

The pea fall-sowing allows extension of the growth period from germination to flowering, resulting in higher biomass yields and increased number of seeds (Uzun and Acikgoz 1998). On the other hand, when pea is sown during fall, it grows and develops in sub-optimal temperature conditions and, in order to survive winter, it needs to withstand the periods of frost, varying in length and intensity in diverse regions. The dissection of the genetic determinism of the winter hardiness in pea is of substantial importance for pea breeders developing fall-sowing pea varieties. Generally, there are three basic adaptation mechanisms in plants regarding winter hardiness: (1) vernalization response, (2) photoperiod sensitivity and (3) cold acclimation. In pea, the second and the third mechanisms are involved in its winter hardiness. Delayed floral initiation helps some forage pea genotypes to escape the main winter freezing periods, as susceptibility to frost increases during the transition to the reproductive state (Lejeune-Hénaut et al. 1999). Numerous studies describe the physiological and phenological effects of the main loci governing the transition to flowering in pea, such as *LF* and *HR*, known to delay floral initiation of autumn-sown peas until a longer day length is reached in the following spring (Lejeune-Hénaut and Delbreil 2009). The oldest winter pea cultivars have the dominant allele, *HR*, although some bear *hr* (Bourion et al. 2002).

Regarding the fact that frost sensitivity appears after floral initiation and that the flowering period is a crucial developmental stage for yield elaboration in pea, the ideotype of the fall-sowing pea cultivars should initiate its flower primordia late enough to escape frosts and should flower early enough to escape drought and heat stresses during late spring (Wenden et al. 2007). In a study carried at INRA Mons, a particular attention was paid to flowering genes in pea, with the assistance of numerous ones describing the physiological and phonological effects of the main loci governing the transition to flowering (Weller et al. 1997). Among these loci related to flowering in pea, *LF* (Late Flowering) and *HR* (High Response to the photoperiod) could be regarded as promising targets, since they control respectively the intrinsic earliness for the beginning of flowering and the strength of the photoperiodic response. In details, the dominant allele *HR* delays the floral initiation in fall-sown peas until a longer day length is reached during the following spring, as well as more favorable temperature conditions. An approach by quantitative genetics was used to check the genetic linkage between the *HR* locus and frost tolerance in pea. A multi-location field trial targeting the frost tolerance was carried out with a population of recombinant inbred lines (RIL) derived by the hybridization of one frost tolerant line, Champagne, and one frost susceptible line, Terese. Champagne also has the dominant allele *HR*, being characterized by a delayed floral initiation under short days in comparison to Terese carrying the recessive allele. The colocalization of *HR* with the major frost tolerance quantitative trait loci (QTL), as was identified within this population, confirmed that this locus plays an important role in the control of frost tolerance in pea (Lejeune-Hénaut et al. 2008). After that, a marker-assisted breeding project aiming at associating the dominant allele at the *HR* locus with the earliest alleles known at the *LF* locus was performed.

A genetic study of cold acclimation in INRA Mons relied on the complementary trials carried out in the field and in controlled conditions. Apart from the *HR* QTL stated above, two other QTL for frost tolerance were consistently mapped on linkage groups (LG) V and VI both for frost damage monitored in the field and in controlled conditions. Several key parameters potentially involved in the cold acclimation in pea, such as electrolyte leakage, concentration of sugars and activity of RuBisCO, were assessed in the resulting mapping population. Some of the detected physiological QTL collocated with the QTL of winter frost damages.

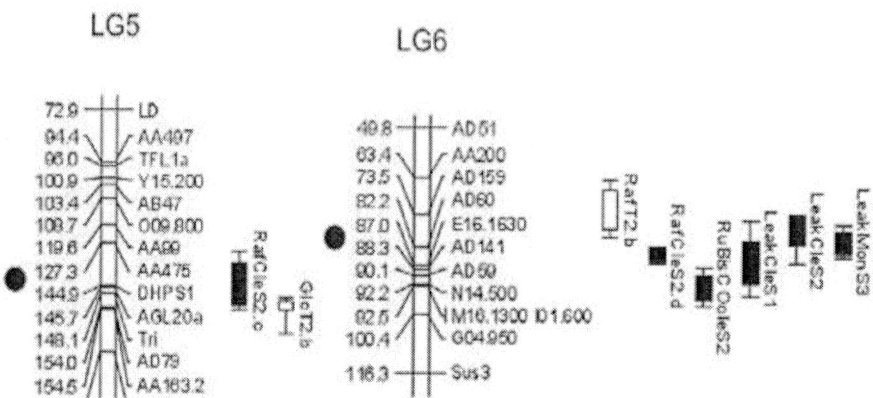

Figure 1. Metabolites, physiological and morphological QTL on LG5 and LG6 in pea. Field data QTL (*black boxes*) in Clermont-Ferrand or Mons are identified by Cle or Mon, respectively. Controlled chamber QTL are represented with *white boxes*. *T2* sampling after 10 days of cold acclimation, *S1,S2* sampling during cold acclimation in the field, *Leak* electrolyte leakage, *Raf* concentration of raffinose, *Glc* concentration of glucose, *RuBisCO* activity of RuBis-CO. *Black circles*, on the *left*, indicate the position of winter frost damage QTL detected within the whole population of recombinant inbred lines from field collected data (Dumont et al. 2009).

Particularly, there were observed three raffinose QTL on LGV and LGVI, one QTL of glucose concentration on LGV and one QTL of RuBisCO activity on LGVI (Figure1).

Additionally, the protein quantitative loci (PQL) were mapped. Among the 22 PQL colocating with a QTL of winter frost damages on LGV, there were four that could be attributed to carbohydrate metabolism and colocated with QTL of glucose and raffinose concentrations. Summarized, these results suggest that carbohydrate metabolism might play a major role in cold acclimation of pea. The QTL for electrolyte leakage measured in field experiments were detected on LGVI. The electrolyte leakage is the physical measurement of membrane damage, used for screening for tolerance to frost damages quantitatively and objectively.

Delayed floral initiation helps some forage pea genotypes to escape the main winter freezing periods, as susceptibility to frost increases during the transition to the reproductive state (Lejeune-Hénaut et al. 1999). Numerous studies describe the physiological and phenological effects of the main loci governing the transition to flowering in pea, such as *Lf* and *Hr*, known to delay floral initiation of autumn-sown peas until a longer day length is reached in the following spring (Lejeune-Hénaut and Delbreil 2009). The oldest winter pea cultivars carry the dominant allele, *Hr*, although some bear *hr* (Bourion et al. 2002). A study of one population of recombinant inbred lines (RILs) allowed detection of six quantitative trait loci (QTL) for frost tolerance, which is in agreement with an oligogenic determinism of frost tolerance in pea. In this population, the most explanatory QTL was found to colocalize with the *Hr* locus (Lejeune-Hénaut et al. 2008). Further studies in the same genetic background gave an insight in the genetic determinism of physiological traits potentially involved in cold acclimation, showing for example the colocalization of QTLs for raffinose concentration or RuBisCO activity with QTLs for frost tolerance on linkage groups 5 and 6 (Dumont et al. 2009).

FALL-SOWING PEA FOR FORAGE PRODUCTION

Breeding winter forage pea emphasizes the development of the lines with satisfying tolerance to low temperatures and more prominent earliness, with non-decreased forage yields and quality. Such line should be able to be cut earlier and leave more time for sowing the succeeding late-spring crops such as silage corn, forage sorghum or Sudan grass.

In a small-plot trial, carried out at the Rimski Šančevi Experiment Field of the Institute of Field and Vegetable Crops, Novi Sad, Serbia, from the autumn 2002 to the summer 2005 and on a slightly carbonated chernozem soil, five winter and five spring cultivars of forage pea of diverse geographical origin were included (Mihailović et al 2007). The winter cultivars were sown in early October, while the spring ones were sown in early March, with a crop density of between 110 plants m^{-2} and 120 plants m^{-2}. All ten cultivars were cut in the stages of full flowering and beginning of formation of pods, that is, in late May and early June, respectively.

The difference between the two groups in average values of yield of green forage per unit area (30.1 t ha^{-1} in the winter cultivars and 27.7 t ha^{-1} in the spring cultivars) was not significant (Table 1). It is apparent that some of the winter cultivars were characterised by high yields of green forage per both plant and area unit, such as NS-Dunav (27.01 g $plant^{-1}$ and 34.1 t ha^{-1}), NS-Pionir (24.85 g $plant^{-1}$ and 30.7 t ha^{-1}) and Osječki Zeleni (21.32 g $plant^{-1}$ and 30.5 t ha^{-1}), as a consequence of prominent winter hardiness and non-significantly reduced stand density. The yield of hay followed the same trend as green forage yield.

Table 1. Agronomic characteristics of winter and spring cultivars of forage pea included in the trial at Rimski Šančevi from 2002 to 2005 (Mihailović et al. 2007b)

Season	Cultivar	Plant height (cm)	Number of inter-nodes (plant^{-1})	Yield of green forage (g plant^{-1})	Yield of green forage (t ha^{-1})	Yield of hay (g plant^{-1})	Yield of hay (t ha^{-1})
Winter	NS-Pionir	140	21.3	24.85	30.7	5.54	6.8
	Champagne	124	20.7	18.10	32.6	5.17	8.3
	Mir	108	16.3	24.33	22.5	4.30	4.0
	NS-Dunav	147	22.0	27.01	34.1	6.26	8.9
	Osječki Zeleni	140	23.7	21.32	30.5	6.05	8.7
	Average	132	20.8	23.12	30.1	5.46	7.3
Spring	NS-Lim	94	19.3	17.52	27.5	2.96	6.5
	P-824	88	19.3	16.84	25.6	3.91	5.9
	NS-Junior	101	19.7	22.73	30.9	2.92	7.1
	Nadja	101	17.3	11.64	23.5	2.86	5.8
	Poneka	99	22.0	21.44	31.0	4.28	6.2
	Average	97	19.5	18.03	27.7	3.38	6.3
LSD	0.05	25	4.5	4.7	5.0	1.8	1.2
	0.01	37	5.9	6.2	9.4	2.5	2.0

Table 2. Mean values of yield (t ha^{-1}) of fresh weight (FW) and hay (H) of forage pea and vetch and their mixtures with small grains (Mihailović et al. 2004)

Cultivar	Year	Single		With wheat		With barley		With oats		Stand mean	
		FW	H	FW	H	FW	H	FW	H	FW	H
Winter pea (NS Pionir)	2000	20.9	4.4	25.2	4.3	48.2	8.7	33.5	6.0	38.2	7.2
	2001	47.3	10.2	43.9	7.5	53.4	9.6	47.7	8.6		
	2002	29.1	6.3	38.1	6.5	35.1	6.3	36.7	6.6		
	mean	32.4	7.0	35.7	6.4	45.6	8.2	39.3	7.1		
Winter pea (NS Dunav)	2000	23.9	5.1	30.0	5.1	46.1	8.3	33.3	6.0	38.7	7.2
	2001	48.1	10.6	46.7	7.9	52.8	9.5	46.8	8.4		
	2002	28.7	6.3	35.1	6.0	36.3	6.5	37.2	6.7		
	mean	33.5	7.4	37.3	6.3	45.0	8.1	39.1	7.0		
LSD	5%	FW	A 1.8	B 2.1	AB 3.0		H	A 0.4	B 0.5	AB 1.4	
	1%		A 2.6	B 5.7	AB 6.8			A 0.5	B 0.7	AB 3.2	

Table 3. Average values of green forage yield (t ha^{-1}), LER$_{GFY}$, forage dry matter yield (t ha^{-1}) and LER$_{FDMY}$ in the mutual intercrops of pea cultivars with different leaf types at Rimski Šančevi during 2008-2010 (Ćupina et al., 2010)

Season	Treatment	Green forage yield of supporting component	Green forage yield of supported component	Total green forage yield	LER$_{GFY}$
Winter	Dove, pure stand	32.4	0.0	32.4	1.00
	Frijaune, pure stand	0.0	30.8	30.8	1.00
	Dove +Frijaune	23.2	11.0	34.2	1.09
Spring	Jezero, pure stand	31.3	0.0	31.3	1.00
	Javor, pure stand	0.0	30.3	30.3	1.00
	Jezero + Javor	16.4	17.5	33.8	1.11
$P < 0.05$		3.7			0.08
Season	Treatment	Forage dry matter yield of supporting component	Forage dry matter yield of supported component	Total forage dry matter yield	LER$_{FDMY}$
Winter	Dove, pure stand	6.8	0.0	6.8	1.00
	Frijaune, pure stand	0.0	7.8	7.8	1.00
	Dove +Frijaune	5.1	3.0	8.1	1.13
Spring	Jezero, pure stand	6.3	0.0	6.3	1.00
	Javor, pure stand	0.0	6.4	6.4	1.00
	Jezero + Javor	2.9	3.6	6.5	1.03
$P < 0.05$		0.8			0.08

The difference between the two groups in average values of yield of green forage per area unit, with 7.3 t ha^{-1} in the winter cultivars and 6.3 t ha^{-1} in the spring cultivars, also was not significant. Among the winter cultivars, the highest yields of hay were in NS-Dunav (6.26 g plant^{-1} and 8.9 t ha^{-1}) and Osječki Zeleni (6.05 g plant^{-1} and 8.7 t ha^{-1}).

In many regions of Europe and North America, fall-sowing pea is cultivated in intercrops with cereals, especially wheat, oat or barley. In a trial carried out at Novi Sad (Table 2) the single stand of fall-sown forage pea had the lowest yield of fresh weight and hay, while the combination of legume and barley in both cases gave the highest yields. The cultivar NS Dunav had the highest yield of fresh weight (38.7 t ha^{-1}), and there was no significant difference between it and the other pea cultivar NS-Pionir. Yields of hay of the two winter pea cultivars were statistically very similar to each other. NS Pionir had the highest yield in combination with barley - 45.6 t ha^{-1} of biomass and 8.2 t ha^{-1} of hay. As a single-stand, it had lower yield of biomass (32.4 t ha^{-1}), and combined with wheat it had the lowest yield of hay (6.4 t ha^{-1}). NS Dunav showed the same behaviour: the highest yields of both biomass and hay with barley (45.0 and 8.1 t ha^{-1}) and the lowest yield of biomass as single-stand (33.5 t ha^{-1}) and of hay with wheat (6.3 t ha^{-1}).

A two-year trial with autumn-sown and spring-sown intercrops of semi-leafless and normal-leafed peas resulted in an economic justification of all four possible combinations, with both Land Equivalent Ratio (LER) values higher than 1 (Ćupina et al. 2010, Table 3). Unpublished preliminary data on the performance of the intercrops of autumn-sown semi-leafless pea with bitter vetch and spring-sown semi-leafless pea with lentil show that the former had somewhat lower economic return in comparison with sole crops, with LER$_{FDMY}$ value of 0.91, while the latter surpassed the pure stands of both crops, with LER$_{FDMY}$ of 1.09. Another goal of the annual feed legume breeding is to find novel uses of the most widely distributed crops. Following this strategy, pre-breeding of the winter dual-purpose pea has begun, resulting in the development of the lines with great potential for both forage and grain yields, such as the cultivar Cer, with about 30 t ha-1 of green forage, 7 t ha-1 of forage dry matter and more than 3100 kg ha-1 of grain.

FALL-SOWING PEA FOR GRAIN PRODUCTION

Breeding spring protein pea is aimed mainly at the selection of the lines with increased earliness and with afila leaf type, since it proved to be at least equal in grain yields in comparison to the cultivars with normal leaf type, while with significantly increased standing ability (Mihailović and Mikić 2004). Apart from this morphological feature, there are others changes in plant architecture, such short internodes and determinate stem growth that are also incorporated in new lines (Mikić et al. 2006). On the other hand, fall-sown dry pea cultivars are becoming more and more in use in many temperate regions in both Europe, especially France, and the USA (McPhee and Muehlbauer 2007). The screening of the winter dry pea accessions from the Novi Sad collection, mostly of French and Bulgarian origin, revealed that some of them had considerable potential for tolerance to low temperatures and grain production. The French lines of recent release, such as Dove and 5174, combined the majority of the desirable characteristics, such as earliness, low winter mortality (Annicchiarico and Iannucci 2008) and high grain yield.

The Serbian programme on fall-sown pea for dry grain production was established in 2004. The first results of the evaluation of the tolerance to low temperatures in several genotypes of diverse geographic origin were encouraging (Mikić et al. 2007). A small-plot trial was carried out on a chernozem soil at the Experiment Field of the Institute of Field and Vegetable Crops at Rimski Šančevi. The trial included six French and three Bulgarian winter protein pea cultivars, namely Champagne, Frilène, Frijaune, Dove, 5105, 5174, Vesela, No. 11 and Drujba, with the Serbian spring protein pea cultivar Javor as a control, being one of the most widely distributed protein pea cultivars in the country. All nine cultivars were sown by hand in early October, at a crop density of about 135 viable seeds m-2, while the control cultivar was sown in early March.

In average, the cultivar 5105 had the earliest date of beginning of flowering (April 16), while the control cultivar Javor had the latest date of beginning of flowering (May 21). The same genotype had the earliest date of harvest (June 11), while the control cultivar Javor had the latest date of harvest (July 1), closely followed by the cultivar Champagne (June 30). The cultivars Dove and 5105 had the highest winter survival coefficients, with 0.93 each in the year of 2004/05 and 0.92 each in the year of 2005/06, while the cultivars Frilène and Vesela had the lowest winter survival coefficients, with 0.16 and 0.21 in the year of 2004/05 and 0.29 and 0.28 in the year of 2005/06 (Table 5).

Table 4. Dates of beginning of flowering and harvest in nine winter protein pea cultivars and a control cultivar (c) in 2004/05 to 2005/06 at Rimski Šančevi (Mikić et al. 2007)

Cultivar	Date of beginning of flowering			Date of harvest		
	2004/05	2005/06	Average	2004/05	2005/06	Average
Champagne	May 10	May 8	May 9	July 1	June 30	June 30
Frilène	April 29	April 18	April 23	June 12	June 17	June 14
Frijaune	April 28	April 18	April 23	June 12	June 17	June 14
Dove	April 27	April 17	April 22	June 17	June 17	June 17
5105	April 20	April 12	April 16	June 9	June 14	June 11
5174	April 23	April 15	April 19	June 20	June 18	June 19
Vesela	April 27	April 20	April 23	June 13	June 18	June 15
No. 11	April 29	April 20	April 24	June 11	June 18	June 14
Drujba	April 30	April 24	April 27	June 24	June 22	June 23
Javor (c)	May 22	May 20	May 21	July 5	June 29	July 1

Table 5. Number of plants before winter (NPW), number of plants before harvest (NPH) and winter surviving coefficient in nine winter protein pea cultivars and a control cultivar (c) in 2004/05 to 2005/06 at Rimski Šančevi (Mikić et al. 2007)

Year Cultivar	NPW (m^{-2})			NPH (m^{-2})			WSC		
	2004/05	2005/06	2004/06	2004/05	2005/06	2004/06	2004/05	2005/06	2004/06
Champagne	131	128	130	107	112	110	0.82	0.87	0.85
Frilène	124	128	126	20	37	29	0.16	0.29	0.23
Frijaune	124	129	127	44	48	46	0.36	0.37	0.37
Dove	132	130	131	123	120	122	0.93	0.92	0.93
5105	129	132	130	120	121	120	0.93	0.92	0.93
5174	129	130	130	106	119	113	0.82	0.92	0.87
Vesela	124	124	124	26	34	30	0.21	0.28	0.24
No. 11	115	124	120	50	50	50	0.44	0.40	0.42
Drujba	127	123	125	74	69	72	0.58	0.56	0.57
Javor (c)	-	-	-	104	112	108	-	-	-
LSD$_{0.05}$	6			7			0.09		
LSD$_{0.01}$	8			10			0.12		

Table 6. Mean values of plant height, grain yield components and grain yield for four dry pea types at Rimski Šančevi from 2004 to 2007 (Mihailović et al. 2008)

Type	Plant height (cm)	Number of fertile nodes (plant^{-1})	Number of pods (plant^{-1})	Number of grains (plant^{-1})	Thousand grain mass (g)	Grain yield (kg ha^{-1})
Winter, conventional	107	7.2	9.8	38.2	117	3348
Winter, semi-leafless	50	3.6	6.6	26.1	206	5236
Spring, conventional	59	3.9	6.3	22.5	236	5215
Spring, semi-leafless	61	3.6	6.0	25.7	233	5252
$LSD_{0.05}$	21	2.6	2.2	8.9	37	943
$LSD_{0.01}$	30	3.6	3.0	12.3	52	1304

Further research was focused on the agronomic performance of the fall-sown dry pea cultivars with different leaf types (Mikić et al. 2011b), namely comparing conventional with semi-leafless ones that have normally developed stipules (*STST*) and leaflets transformed into tendrils (*afaf TLTL*). With a regular sowing in early October, the semi-leafless fall-sown dry pea cultivars produce grain yields comparable to those of spring cultivars (Table 6), thus confirming that the growing of fall-sown dry pea cultivars with reduced height, short internodes, semi-leafless type, white flowers and light-coloured testa (McPhee et al. 2007), is possible in many temperate agroecological conditions of northern hemisphere, deserving full attention by both breeders and farmers of the strategic advantages of autumn-sown dry pea cultivars of recent release is their significantly improved earliness (Figure 2). The Serbian fall-sown dry pea cultivar Mraz, newly registered in Serbia and developed from hybrid populations of fall-sown material of French and Serbian origin, is regularly at least one week earlier than winter barley, providing many farmers with the novel opportunity of not having to choose between pea and cereals, since many have only one combine harvester available and give priority to their cereals. Furthermore, winter dry pea cultivars may have increased grain dry matter crude protein content (UNIP and ITCF 1995), and it is possible to merge winter hardiness and low content of anti-nutritional factors (Ney and Duc 1997). Low thousand seed weight, not exceeding 200 g, and a population density of 75-80 plants m^{-2} (Knott and Belcher 1998) provide inexpensive sowing. Most recently, the first fall-sown semi-leafless dry pea cultivars with delayed flowering (*HR*) were released, such as Jeronimo in France, assessing the possibility of earlier sowing during the fall.

CONCLUSIONS

The achieved results in the breeding and the cultivation of fall-sown pea in contrasting temperate regions such as France, USA or Serbia confirm that it could be one of the least expensive and most efficient ways to decrease the unpredictable and destroying effects of spring droughts and other manifestations of climatic changes on protein-rich crops such as pea. They also establish a solid basis for the anticipation that the existence of high-yielding, early and winter hardy fall-sown dry pea cultivars will increase the total area under grain legumes, especially in Europe, and thus contribute to a significant increase of the protein needed for ever demanding animal husbandry.

ACKNOWLEDGMENTS

The projects TR-31016 and TR-31024 of the Ministry of Education and Science of the Republic of Serbia and the project LEG-HIVER within the programme *Pavle Savić* of the bilateral cooperation between France and Serbia.

Figure 2. The latest Serbian winter dry pea line L-574, registered in November 2010 under the name of the cultivar Mraz (right), in comparison to the widely used Serbian spring-sown dry pea cultivar Jezero (left), photographed on the same day, May 12, 2009 (Mikić et al. 2011a).

REFERENCES

Annicchiarico P, Iannucci A (2008) Adaptation strategy, germplasm type and adaptive traits for field pea improvement in Italy based on variety responses across climatically contrasting environments. *Field Crop Res* 108:133–142.

Bourion V, Fouilloux G, Le Signor C, Lejeune-Hénaut I (2002) Genetic studies of selection criteria for productive and stable peas. *Euphytica* 127: 261–273.

Bourion V, Lejeune-Hénaut I, Munier-Jolain N, Salon C (2003) Cold acclimation of winter and spring peas: carbon partitioning as affected by light intensity. *Eur J Agron* 19:535-548.

Burstin J (2009) New challenges and opportunities for pea. *Grain Legum* 52:4.

Carrouée B (1993) Different types of peas: to clarify a complex status. *Grain Legum* 3: 26-27.

Ćupina B, Krstić Đ, Antanasović S, Erić P, Pejić B, Mikić A, Mihailović V (2010) Potential of the intercrops of normal-leafed and semi-leafless pea cultivars for forage production. *Pisum Genet*, 42, 11-14.

Duc G (1997) Faba bean (*Vicia faba* L.). *Field Crop Res* 53:99-109.

Dumont E, Fontaine V, Vuylsteker C, Sellier H, Bodèle S, Voedts N, Devaux R, Frise M, Avia K, Hilbert JL, Bahrman N, Hanocq E, Lejeune-Hénaut I, Delbreil B (2009) Comparative studies of frost tolerance QTL in *Pisum sativum* L. under field and controlled conditions using physiological and phenotypical traits. *Theor. Appl. Genet.* 118:1661-1571.

FAOSTAT (2007): ProdSTAT: Crops. FAO Statistical Databases (FAOSTAT), Food and Agriculture Organization of the United Nations (FAO), *http://faostat.fao.org.*

Knott CM, Belcher SJ (1998). Optimum sowing dates and plant populations for winter peas (*Pisum sativum*). *J Agric Sci* 131:449-454.

Lejeune-Henaut I, Delbreil B (2009) Genetics of winterhardiness in pea. *Grain Legum* 52:7-8.

Lejeune-Hénaut I, Bourion V, Etévé G, Cunot E, Delhaye K, Desmyter C (1999) Floral initiation in field-grown forage peas is delayed to a greater extent by short photoperiods, than in other types of European varieties. *Euphytica* 109:201-211.

Lejeune-Hénaut I, Hanocq E, Béthencourt L, Fontaine V, Delbreil B, Morin J, Petit A, Devaux R, Boilleau M, Stempniak J-J, Thomas M, Lainé A-L, Foucher F, Baranger A, Burstin, Rameau C, Giauffret C (2008) The flowering locus Hr colocalizes with a major QTL affecting winter frost tolerance in Pisum sativum L. *Theor Appl Genet* 116:1105-1116.

Maxted N, Ambrose M (2000) Peas (Pisum L.). In: Maxted, N., Bennet, S. (eds) Plant genetic resources of legumes in the Mediterranean, Kluwer Academic Publishers, Dordrecht, The Netherlands, 181-190.

McPhee KE, Muehlbauer FJ (2007) Registration of 'Specter' Winter Feed Pea. *J Plant Regist* 1:118–119.

McPhee KE, Chen CC, Wichman DM, Muehlbauer FJ (2007) Registration of 'Windham' Winter Feed Pea. *J Plant Regist* 1:117–118.

Mihailović V, Mikić A (2004) Leaf type and grain yield in forage pea. *Genet Belg*, 36, 1, 31-38.

Mihailović V, Mikić A, Ćupina B (2004a) Botanical and agronomic classification of fodder pea (*Pisum sativum* L.). *Acta Agric Serb* IX:17(special issue):61-65.

Mihailović V, Erić P, Mikić A. (2004) Growing peas and vetches for forage in Serbia and Montenegro. *Grassl Sci Eur* 9:457-459.

Mihailović V, Ćupina B, Mikić A, Katić S, Karagić Đ (2007a) Relationships between agronomic characteristics in dual-purpose pea. *Book of Abstracts of the 6th European Conference on Grain Legumes Integrating Legume Biology for Sustainable Agriculture*, Lisbon, Portugal, 12-16 November 2007, 116.

Mihailović V, Mikić A, Vasiljević S, Milić D, Ćupina B, Erić P (2007b) Agronomic characteristics of winter and spring cultivars of forage pea (*Pisum sativum* L.). Proceedings of the XXVI EUCARPIA Fodder Crops and Amenity Grasses Section and XVI Medicago spp. *Group Joint Meeting Breeding and Seed Production for Conventional and Organic Agriculture,* Perugia, Italy, 3-7 September 2006, 182-184.

Mihailović V, Ellis THN, Duc G, Lejeune-Hénaut I, Étévé G, Angelova S, Mikić A, Ćupina B (2008) Grain yield in winter and spring protein pea cultivars (*Pisum sativum* L.) with normal and afila leaf type. *Proceedings of the International Conference Conventional and Molecular Breeding of Field and Vegetable Crops, Novi Sad*, Serbia, 24-27 November 2008, 443-446.

Mihailović V, Mikić A (2010) Novel directions of breeding annual feed legumes in Serbia. *Proceedings, XII International Symposium on Forage Crops of Republic of Serbia*, Kruševac, Serbia, 26-28 May 2010, 1, 81-90.

Mikić A, Mihailović V, Katić S, Karagić Milić D (2003) Protein pea grain - a quality fodder. *Biotechnol Anim Husb* 19:5-6:465-471.

Mikić A, Mihailović V, Milić D, Vasiljević S, Katić S, Ćupina B (2006) The role of *af*, *det* and *le* genes in increasing grain yield of feed pea (*Pisum sativum* L.) in Serbia and Montenegro. *Handbook and Abstracts of the 3rd International Conference on Legume Genomics and Genetics Genes to Crops,* Brisbane, Australia, 9-13 April 2006, 109.

Mikić A, Mihailović V, Duc G, Ćupina B, Étévé G, Lejeune-Hénaut I, Mikić V (2007) Evaluation of winter protein pea cultivars in the conditions of Serbia. *Ratar Povrt / Field Veg Crop Res* 44:II:107-112.

Mikić A, Perić V, Đorđević V, Srebrić M, Mihailović V (2009) Anti-nutritional factors in some grain legumes. *Biotechnol Anim Husb* 25:1181-1188.

Mikić A, Mihailović V, Ćupina B, Đorđević V, Milić D, Duc G, Stoddard FL, Lejeune-Hénaut I, Marget P, Hanocq E (2011a) Achievements in breeding autumn-sown annual legumes for temperate regions with emphasis on the continental Balkans. *Euphytica* 180:57-67.

Mikić A, Mihailović V, Ćupina B, Kosev V, Warkentin T, McPhee K, Ambrose M, Hofer J, Ellis N (2011b) Genetic background and agronomic value of leaf types in pea (Pisum sativum). *Ratar povrt / Field Veg Crop Res* 48:275-284.

Nemecek T, von Richthofen JS, Dubois G, Casta P, Charles R, Pahl H (2008) Environmental impact of introducing grain legumes into European crop rotations. *Eur J Agron* 28:380–393.

Ney B, Duc G (1997) Potential and problems with winter sowing of food legumes in northern Europe. *Proceedings, AEP Workshop Problems and Prospects for Winter Sowing of Grain Legumes in Europe*, Dijon, France, 3-4 December 1996, pp. 35-42.

Shereena J, Salim N (2006) Chilling tolerance in Pisum sativum L. seeds: an ecological adaptation. *Asian J Plant Sci* 5:1047-1050.

Sincik M, Bilgili U, Uzun A, Açikgöz E (2004) Effect of low temperatures on the germination of different field pea genotypes. *Seed Sci Technol* 32: 331-339.

UNIP, ITCF (1995): Peas - Utilisation in Animal Feeding. UNIP –ITCF, Paris.

Uzun A, Acikgoz E (1998) Effect of sowing season and seeding rate on the morphological traits and yields in pea cultivars of differing leaf types. *J Agron Crop Sci* 181:215-222.

Weller JL, Reid JB, Taylor SA, Murfet IC (1997) The genetic control of flowering in pea. *Trends Plant Sci* 2:411-418.

Wenden B, Rameau C, Lejeune-Henaut I (2007) Manipulating flowering time. *Grain Legum* 49:7-8.

Yordanov I, Georgieva K, Tsonev Ts, Velikova V (1996) Effect of cold hardening on some photosynthetic characteristics of pea (Pisum sativum L., cv. Ran 1) plants. *Bulg J Plant Physiol* 22(1–2):13–21.

In: Peas   ISBN: 978-1-61942-866-9
Editors: A. Comstock and B. Lothrop   © 2012 Nova Science Publishers, Inc.

Chapter 5

PEAS (*PISUM SATIVUM*): CHEMICAL, MOLECULAR STRUCTURAL, BIODEGRADATION AND NUTRITIONAL CHARACTERIZATION

Peiqiang Yu[*]
College of Agriculture and Bioresources, The University
of Saskatchewan, Saskatoon, SK, Canada

1. INTRODUCTION

The Latin name for pea seeds is *Pisum sativum*. The peas are also called marrowfat pea, small blue pea, dry pea, arveja, muttar etc (Petterson and Mackintosh, 1994; Goelema, 1999; Yu et al., 2002). Peas have attracted attention as components of food and feed for human and animals as protein and energy supplements mainly because they usually have a particularly high protein and starch contents. Peas are also as one of health food for human due to fact that they are high in fiber, vitamins, minerals and lutein etc. Although pea seed peptide fractions have less ability to scavenge free radicals than glutathione, greater ability to chelate metals and inhibit linoleic acid oxidation (http://en.wikipedia.org/wiki/Pea#Nutritional_value; Pownall et al., 2010).

---

[*]Corresponding author contact: Tel: (306) 966 4132, E-mail: peiqiang.yu@usask.ca

Peas have a diversity of varietiesare well suited to the various ecological and climatic conditions in many countries with high seed yields (Cerning-Beroard and Filiatre, 1977; Ensminger and Olentine, 1978; Golema, 1999; Yu et al., 2002). However, the use of peas in ruminants (for example in dairy cows) is limited and the utilization is inefficient mainly due to the fact that soluble or rapidly degradable protein content and the rate of degradation are too high (Van Straalen and Tamminga, 1990; Goelema, 1999; Yu et al., 2002). These cause an imbalance between protein breakdown and microbial protein synthesis, resulting in unnecessary N loss from the rumen (Goelema, 1999; Yu et al., 2002). To be more efficiently used, the digestive characteristics should be manipulated through processing or physical, chemical and biological treatments. In this article, the information on chemical, molecular structural, biodegradation and nutritional characteristics of peas for animals are reviewed.

2. STRUCTURAL SPECTRAL FEATURES OF PEAS (*PISUM SATIVUM*) USING MOLECULAR SPECTROSCOPY: COMPARISON AMONG PEAS HULL, WHOLE SEED AND SEED WITHOUT HULL

The molecular structures are closely related to nutrient availability and utilization. The inherent structure profiles and structural changes induced by various processing or physical and chemical treatments can be detected using various molecular spectroscopy such as advanced synchrotron radiation-based SR-IMS technique (Yu, 2004, 2010; Yu et al., 2004), DRIFT (Doiron et al., 2009a, 2009b; Liu and Yu, 2010) and FT/IR-ATR molecular spectroscopy (Yu, 2010; Samadi and Yu, 2011). The following are examples from pea (cv. Homesteader) spectral feature study in my research team in 2011.

In this study, the molecular spectral data of pea whole seeds, pea hull, seed without hull were collected and corrected with the background spectrum using JASCO FT/IR-ATR 4200. The spectra were generated with mid-IR (ca. 4,000-800 cm^{-1}; Figures 1-3) with spectral resolution of 4 cm^{-1}. The FT/IR spectral data was collected using JASCO Spectral Managers and analyzed using OMINIC 7.2 (Spectra-Tech, Madison, WI) software.

2.1. Univariate Molecular Spectral Analysis of Peas

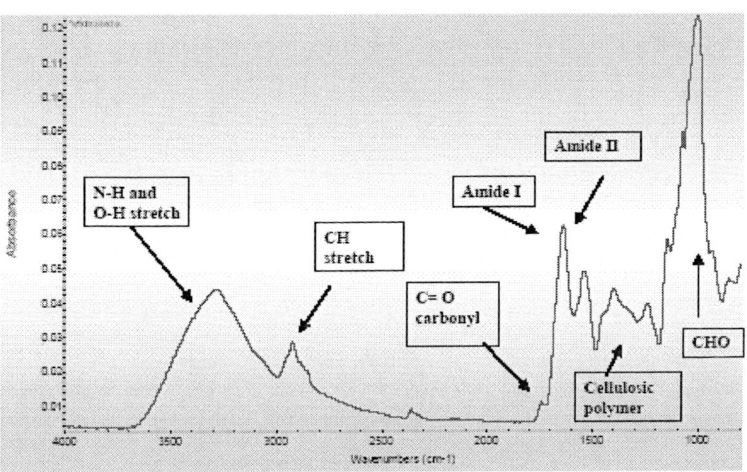

Figure 1. Typical FT/IR molecular spectrum in whole seeds of peas (*Pisum sativum* cv. Homesteader) in the region ca. 4000-800 cm^{-1} showed function groups in legume seeds: N-H and O-H stretch, C-H stretch, amide I and II, C = O carbonyl, CHO and cellulosic compounds.

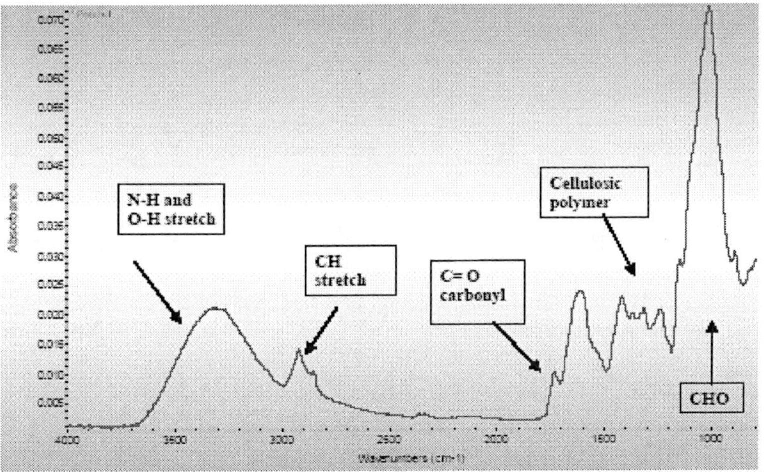

Figure 2. Typical FT/IR molecular spectrum in the hull of peas (*Pisum sativum* cv. Homesteader) in the region ca. 4000-800 cm^{-1} showed function groups in legume seeds: N-H and O-H stretch, C-H stretch, C = O carbonyl, CHO and cellulosic compounds.

Chemical functional groups and their ratios were identified according to various published reports (Kemp, 1991; Himmelbsbach, et al., 1998; Wetzel, et al., 1998; Wetzel, 2001; Miller, 2002; Marinkovic and Chance, 2006). The regions of specific interest in this study included peaks at ca. 1725 (carbonyl C=O ester), ca. 1650 (amide I), ca. 1657 (protein 2nd structure α-helix), ca. 1628 (protein 2nd structure β-sheet), ca, 1550 (amide II), ca. 1515 (aromatic compounds of lignin), 1428, 1371 and 1245 (cellulosic compounds), 1025 (non-structural CHO, starch granules), 1246 (cellulosic compounds), 1160 (CHO), 1150 (CHO), 1080 (CHO), 930 (CHO), 860 (CHO), 3350 (OHandNH stretching), 2960 (CH_3 anti-symmetric), 2929 (CH_2 anti-symmetric), 2877 (CH_3 symmetric) and 2848 cm^{-1} (CH_2 asymmetric) (Yu, 2010; 2011).

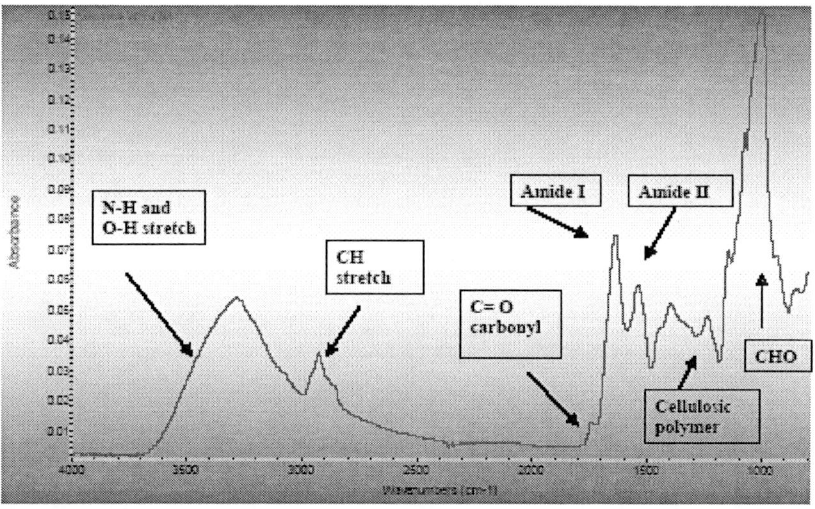

Figure 3. Typical FT/IR molecular spectrum in inside seed without hull of peas (*Pisum sativum* cv. Homesteader) in the region ca. 4000-800 cm^{-1} showed function groups in legume seeds: N-H and O-H stretch, amide I and II, C-H stretch, C = O carbonyl, CHO and cellulosic compounds.

Relative functional groups intensity ratios such as protein 2nd structure α-helix–to-β-sheet ratio, protein amide I-to-starch granule ratio, and anti-symmetric CH_3-to-CH_2 ratio were aslo investigated to show the molecular structural difference by using molecular univariate spectral analysis.

Figures 1-3 showed that the molecular spectral features, pattern and various functional group intensities in whole seed, hull and seed without hull of peas different.

2.2. Multivariate Molecular Spectral Analyses

Two spectral analyses, hierarchical cluster analysis (CLA) and principal components analysis (PCA), were used to compare the original spectra of whole seed, hull and seed without hull of peas to determine underlying structural differences in the spectral information that permits the identification of spectra belonging to different groups. The spectra were used without smoothing and parameterization. No FSD spectra and the 2nd derivative spectra were used.

For CLA and PCA the spectral fingerprint region of ca. 1800-800 cm^{-1}, containing most functional group bands were used. CLA results were presented as dendograms while PCA results were plotted based on the three highest factor scores and plotted as a function of those scores. In each comparison the eigenvector for factor 1 was plotted against that of factor 2 which accounted for over 70 of the variability in the data. Multivariate spectral analyses were performed using Statistica 6.0 software (StatSoft Inc, Tulsa, OK, USA).

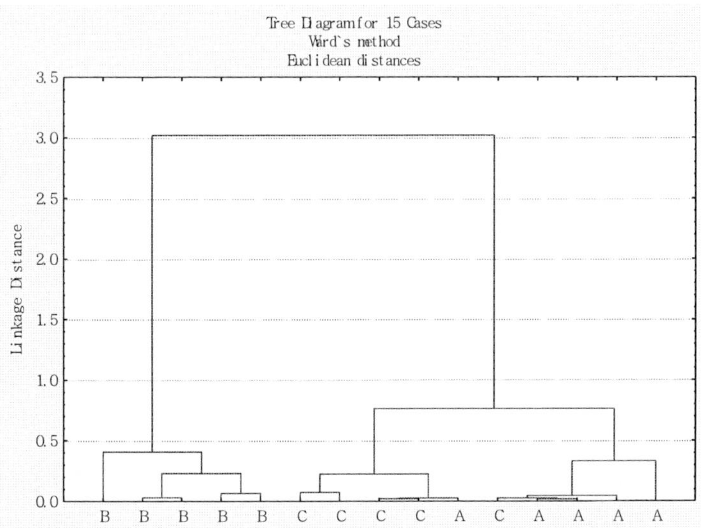

Figure 4. Multivariate molecular spectral analyses for whole seed of peas (code A), pea hull (code B) and pea inside seed without hull (code C): CLA cluster analyses of fingerprint spectrum region ca. 1800-800 cm^{-1}; Distance method: Euclidean; Cluster method: Ward's algorithm).

The results (Figures 4 and 5) show that the multivariate molecular spectral analyses, cluster analysis and principal component analysis of original spectra (without spectral parameterization), distinguished the structural differences between the hull and whole seed or seed without hull in fingerprint spectral region (ca. 1800-800 cm^{-1}). Further study is needed to quantify molecular structural changes in relation to nutrient utilization.

The molecular technique and methodology provide a high potential that could be used to monitor the degree of pea maturity and damage and the fate of organic contaminants and effect of physical, chemical or biological treatment to pea plant and seeds.

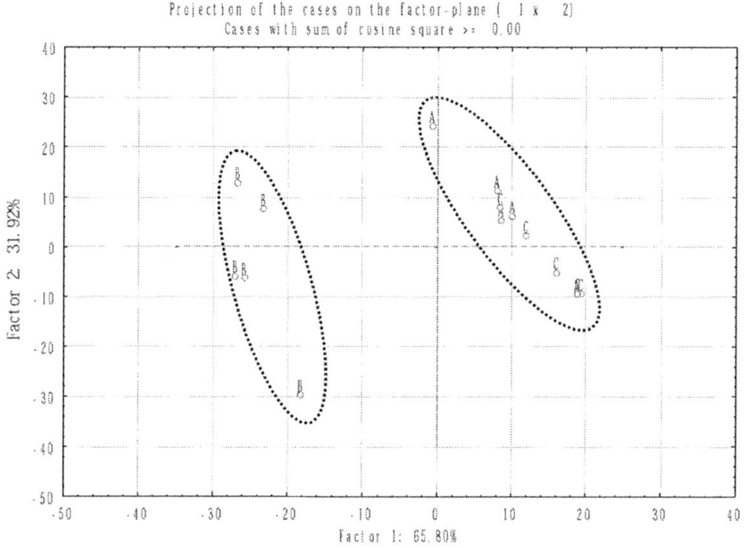

Figure 5. Multivariate molecular spectral analyses for whole seed of peas (code A), pea hull (code B) and pea inside seed without hull (code C): PCA analyses of fingerprint spectrum region ca. 1800-800 cm^{-1}

3. CHEMICAL AND BIODEGRADATION CHARACTERISTICS AND NUTRIENT PROFILES OF PEAS (*PISUM SATIVUM*)

3.1. Unique Chemical Profiles of Peas

Peas (*Pisum sativum*) have unique chemical profiles (Table 1). They contain high protein (ca. 250 g/kg DM) and high starch (ca. 450 g/kg DM). Large of proportion of crude protein is soluble crude protein (>75%). In soluble protein, one third of them are non-protein nitrogen. Peas contain low lignin and low acid detergent insoluble protein (<1%). The unique chemical composition result in unique chemical ratio which have benefit to animals. Peas contain a high ratio of non-structural carbohydrate to structural carbohydrate (6.44) and low ratio of lignin to neutral detergent fibre (0.08) (Table 1).

Table 1. Chemical profiles of peas (*Pisum sativum*)

Items	Values
Basic chemical profile	
Ash (g/kg DM)	33.5
Crude fat (g/kg DM)	11.8
Protein profile	
Crude protein (CP, g/kg DM)	249.8
Soluble crude protein (SCP, g/kg DM)	196.8
Soluble crude protein (SCP, g/kg CP)	788.0
Non-protein nitrogen (NPN, g/kg SCP	317.3
Non-protein nitrogen (NPN, g/kg CP)	250.0
Non-protein nitrogen (NPN, g/kg DM)	62.5
Acid detergent insoluble protein (ADIP, g/kg DM)	6.0
Acid detergent insoluble protein (ADIP, g/kg CP)	24.0
Neutral detergent insoluble protein (NDIP, g/kg DM)	23.0
Neutral detergent insoluble protein (NDIP, g/kg CP)	92.1
Carbohydrate profile	
Carbohydrate (CHO, g/kg DM)	704.9
Non-structural carbohydrate (NSC, g/kg DM)	630.1
Starch (g/kg DM)	455.0
Starch (g/kg NSC)	722.1
Neutral detergent fibre (g/kg DM)	97.8
Acid detergent fibre (g/kg DM)	71.9
Acid detergent lignin (g/kg DM)	7.6
Acid detergent lignin (g/kg NDF)	77.7
Hemicullulose (g/kg DM)	25.9
Cellulose (g/kg DM)	64.3
Unique chemical ratio	
Ratio of crude protein to carbohydrate	0.35
Ratio of non-structural carbohydrate to structural carbohydrate	6.44
Ratio of lignin to neutral detergent fibre	0.08
Ratio of acid detergent fibre to neutral detergent fibre	0.74
Ratio of cellulose to hemicellulose	2.48

Basic data used for calculation from NRC (2001) and Goelema (1999)

3.2. Protein and Carbohydrate Subfractions Associated with Degradation Characteristics in Peas

Peas (*Pisum sativum*) have unique protein and carbohydrate subfractions (Table 2). These fractions are closely related to animal digestive behaviors. According to Cornell Net Carbohydrate and Protein System (CNCPS) (Sniffen el al., 1992), the protein was partitioned into five subfractions: PA, PB1, PB2, PB3, PC. The carbohydrate was portioned into four subfractions: CA, CB1, CB2 and CC. These subfractions are closely related to degradation rate and nutrient availability in ruminants. These fractions are key items that have been used in CPM, CNCPS and new DVE-2007 model diet formulation. The details of each fraction has been reported before as follow:

"Fraction PA is non-protein nitrogen, fraction PB is true protein, and fraction PC is unavailable protein. Fraction PB is further divided into three subfractions (PB1, PB2, and PB3) that are believed to have different rates of degradation; Buffer-insoluble protein minus fraction PB3 is used to estimate fraction PB2. Fraction PB2 is insoluble in buffer but soluble in neutral detergent, while fraction PB3 is insoluble in both buffer and neutral detergent, but is soluble in acid detergent. Fraction PB2 is fermented at a lower rate than buffer-soluble fractions, and some of the PB2 fraction escapes to the lower gut. Fraction PB3 is believed to be more slowly degraded than fractions PB1 and PB2 because of its association with the plant cell wall; a large proportion of PB3 is thus believed to escape fermentation. Fraction PC is the acid detergent insoluble N, which is highly resistant to breakdown, and it is assumed to be unavailable. The relative degradation rates of the five protein fractions are: fractions PA is assumed to be infinity, fraction PB1 is 1.20-4.00 /h, fraction PB2 is 0.03-0.16/h, fraction PB3 is 0.0006-0.0055/h. Fraction PC is considered to be undegradable. The CHO fractions were partitioned, according to the CNCPS system (Sniffen el al., 1992), into: a rapidly degradable fraction (CA) which is composed of fermentable soluble sugars that have a rapid degradation rate of 3.00/h, intermediately degradable fraction (CB1) which is starch and pectin with an intermediate degradation rate of 0.20-0.50/h, a slowly degradable fraction (CB2) which is available cell wall with a slow degradation rate of 0.02-0.10/h, and an unfermentable fraction (CC) which is the unavailable cell wall (Sniffen el al., 1992)."

Pea crude protein contain 73% true protein (Table 2). Fraction PA is 25.0% of CP, PB fractions 72.6% of CP. Fraction PC is very low 2.4% of CP. Among true protein PB fractions, PB1 is 74.1% of true protein, PB2 is 16.5% of true protein and PB3 is 9.3 % of true protein.

In pea carbohydrate fractions, fraction CA is 24.8% of CHO, CB fraction is 75.4% of CHO. Fraction CC is also very low 2.6% of carbohydrate. Among carbohydrate CB fractions, CB1 is 64.6% of carbohydrate and CB2 is 8.0% of carbohydrate (Table 2).

Table 2. Protein and carbohydrate subfractions of peas (*Pisum sativum*) associated with degradation characteristics

Items	Values
Protein subfractions	
Fraction of PA (%CP)	25.0
Fraction of PB1 (%CP)	53.8
Fraction of PB2 (%CP)	12.0
Fraction of PB3 (%CP)	6.8
Fraction of PC (%CP)	2.4
True Protein (TP, % CP)	72.6
Fraction of PB1 (%TP)	74.1
Fraction of PB2 (%TP)	16.5
Fraction of PB3 (%TP)	9.3
Fraction of PA (%DM)	6.3
Fraction of PB1 (%DM)	13.4
Fraction of PB2 (%DM)	3.0
Fraction of PB3 (%DM)	1.7
Fraction of PC (%DM)	0.6
Carbohydrate subfractions	
Fraction of CA (%CHO)	24.8
Fraction of CB1 (%CHO)	64.6
Fraction of CB2 (%CHO)	8.0
Fraction of PC (%CHO)	2.6
Fraction of CA (%DM)	17.5
Fraction of CB1 (%DM)	45.5
Fraction of CB2 (%DM)	5.7
Fraction of PC (%DM)	1.8

3.3. Available Nitrogen and Available Carbohydrate Ratio in Peas

Peas also have unique avaiable nitrogen to available carbohydrate fraction ratios (Table 3). For example, available NA to available CA is 57 g N /kg DM CHO, available NB1 to available CB1 is 47 g N /kg DM CHO and available NB2 to available CB2 is 85 g N /kg DM CHO. These data affect degraded protein balance in rumen (Tamminga et al., 1994; 2007). For optimum microbial protein synthesis, the ratio of rumen available proteins to rumen available carbohydrates is a critical determinant (Yu et al., 2002). Czerkawski (1986) reported that around 25 g N per kg available carbohydrates should be appropriate and sufficient for achieving optimum microbial growth. Tamminga et al. (1990) reported that optimum ratio of carbohydrates (g/kg DM) and protein (g/kg DM) is approximately 5:1 (approximately 33 g N per kg carbohydrates). Peas usually have the higher ratio of available nitrogen to available carbohydrates.

Table 3. Ratio of available protein or nitrogen and available carbohydrate of peas (*Pisum sativum*)

Items	Values
Ratio of available protein to available carbohydrate fractions	
Ratio of PA to CA fractions	0.36
Ratio of PB to CB fractions	0.35
Ratio of PB1 to CB1 fractions	0.30
Ratio of PB2 to CB2 fractions	0.53
Ratio of available nitrogen to carbohydrate (g N/kg CHO)	
Ratio of NA to CA fractions	57.06
Ratio of NB to CB fractions	56.72
Ratio of NB1 to CB1 fractions	47.26
Ratio of NB2 to CB2 fractions	84.75

3.4. Energy Value of Peas

Using the NRC (1996, 2001) chemical summary approach approaches, digestible nutrients (tdCP, tdFA, tdNDF, tdNFC) and the energy value (TDN, DE, ME, NE) of peas can be estimated. Table 4 shows the details total digestible nutrients and energy values of peas. Peas have high TDN value

(>80% DM), Net energy value at a production level is 1.99 Mcal/kg DM for lactation dairy cattle (3x) and 2.15 Mcal/kg DM and 1.47 Mcal/kg DM for beef cattle at maintenance and growth levels, respectively.

Table 4. Digestible nutrient and energy values of peas (*Pisum sativum*) estimated using NRC 1996 and 2001

Items	Values
Digestible nutrients estimated using NRC (% DM)	
tdNDF	3.94
tdNFC	61.75
tdCP	24.27
tdFA	0.18
Digestion of nutrients estimated using NRC 2001	
%dNDF	40.3
%dNFC	98.0
%dCP	97.2
%dFA	15.3
Total digestible nutrient	
TDN1x (% DM)	83.37
Energy value for dairy cattle (Mcal/kg DM)	
DE1x	3.84
DE3x	3.52
ME3x	3.11
NEL3x	1.99
Energy value for beef cattle (Mcal/kg DM)	
ME	3.14
NEm	2.15
NEg	1.47

3.5. In Situ Rumen Degradation Kinetics of Peas (*Pisum Sativum*)

Although peas contain high protein and high starch, the use of peas in ruminants (dairy cows) is limited and the utilization is inefficient under certain condition (Goelema, 1999; Yu et al., 2002). This is because that soluble or rapidly degradable protein content and the rate of degradation are too high (Van Straalen and Tamminga, 1990; Walhain et al., 1992; Aguilera et al., 1992; Goelema, 1999), therefore their effective degradability of protein and starch are high too with 75-88 % of CP and 59-75 % of starch, respectively. Such high values cause an imbalance between protein breakdown and microbial protein synthesis, resulting in unnecessary N loss from the rumen and results less rumen bypass protein (33 to 58 g/kg DM) and starch (107 to 171 g/kg DM).

Table 5. Degradation kinetics of protein and starch in peas

Items	Values
Protein degradation kinetics of peas	
Soluble fraction (% of CP)	39.3 - 62.1
Potentially degradable fraction (% of CP)	37.9 - 60.7
Lag time (h)	0
Degradation rate (%/h)	4.8 - 20.0
Effective degradability of protein (% of CP)	75.1 – 88.0
Rumen undegraded protein (% of CP)	11.8 – 25.0
Effective degraded protein (g/kg DM)	197.4 – 226.5
Rumen undegraded protein (g/kg DM)	33.5 – 58.4
Starch degradation kinetics of peas	
Soluble fraction (% of starch)	13.8 – 47.8
Potentially degradable fraction (% of starch)	52.2 – 86.2
Lag time (h)	0
Degradation rate (%/h)	4.6 – 14.1
Effective degradability of starch (% of starch)	59.1 – 74.9
Rumen undegraded starch (% of starch)	25.1 – 40.9
Effective degraded starch (g/kg DM)	247.2 – 322.5
Rumen undegraded starch (g/kg DM)	106.9 – 170.8

Adapted from Van Straalen and Tamminga (1990), Walhain et al. (1992), Aguilera et al. (1992), Goelema (1999) and summaried by Yu et al. (2002)

4. MANIPULATION OF DIGESTIVE KINETICS OF PEAS (*PISUM SATIVUM*)

To more efficiently use peas, the digestive characteristics of peas should be manipulated through various processes such as physical, chemical and biological treatments or through peas breeding programs to lower degradation extent and rate. Conventional methods to reduce degradation rate and extent of peas in rumen are heat processing which include dry roasting, moisture roasting, micronization, toasting, pressure toasting, extrusion and pelleting

(Goelema, 1999; Yu et al., 2002). However, different methods have different consequence, sometimes produce opposite results.

Goelema (1999) manipulated digestive behavior of protein through pressure toasting at different times (3-30 min) and temperatures from 100-136°C and made peas protein and starch more bypassing the rumen. The pressure toasting reduced the soluble fraction from 62 to 25% of CP, increased the potentially degradable fraction 38 to 75% of CP, reduce the protein degradation rate (%/h) from 4.8 to 2.4 %/h and effective degradability of protein 89 to 56% and increase the rumen undegraded protein from 24 to 54 %. The pressure toasting also changed the starch degradation kinetics, reduce the soluble fraction from 52 to 11% of starch, increased the potentially degradable fraction 48 to 89% of starch, reduced the effective degradability of starch 68 to 50 % of starch and increased the rumen undegraded starch from 32 to 51 % of starch. Walhain et al. (1992) used extrusion (140 to 220°C) to manipulate the digestive behavior. From these studies, It was found that the sensitivity of biopolymers in peas to heating were different between protein and starch. Effects of processing methods to peas were also different on degradation kinetics and nutrient supply. More research need to be done in detecting the sensitity of biopolymers or chemical functional groups to the processing (such as heating).

5. MODELING NUTRIENT SUPPLY FROM PEAS (*PISUM SATIVUM*): DEGRADED PROTEIN BALANCE, MICROBIAL PROTEIN SYNTHESIS AND TRULY ABSORBABLE PROTEIN IN THE SMALL INTESTINE

To predict nutrient supply from peas to ruminant, a Dutch protein evaluation system was applied (Tamminga et al., 1994; 2007). As mentioned by Tamminga et al. (1994 and 2007), "this new system considers the strong elements of other recently developed protein evaluation systems and it also introduces new elements, such as the role of energy balance in protein supply (Tamminga et al., 1994). In this system, each feed has a true protein value, which is truely absorbable protein in the small intestine, composed of the digestible true protein contributed by digestible feed true protein escaping rumen degradation, digestible true microbial protein synthesized in the rumen and a correction for endogenous protein losses in the digestive tract. Each feed also has rumen degraded protein balance, which shows the (im)balance between microbial protein synthesis potentially possible from available rumen degradable protein, and that potentially possible from the energy extracted during anaerobic fermentation in the rumen. When degraded protein balance is positive, it indicates the potential loss of N of a feed from the rumen. When negative, microbial protein synthesis may be potentially impaired because of a shortage of N of a feed in the rumen (Tamminga et al., 1994; 2007)."

Table 6 presents detailed potential nutrient supply from peas to dairy cattle. Briefly, peas provide 49 g/kg DM of the digestible feed true protein escaping rumen degradation, 57 g/kg DM of digestible true microbial protein synthesized in the rumen and 12 g/kg DM of a correction for endogenous protein losses in the digestive tract. Total truly and absorbed peas protein in the intestine is around 94 g/kg DM. The degraded balance is 103 g/kg DM indicating potential protein loss when feed peas alone.

Table 6. Modeling nutrient supply from peas to dairy cattle

Items (g/kg DM)	Values
Total digestible organic matter	821.5
Indigestible organic matter	145.0
Digestible ash	21.8
Digestibility of ash	65%
Indigestible ash	11.7
Indigestible dry matter	156.7
Rumen bypass protein (Dutch model)	58.2
Rumen bypass starch (Dutch model)	158.4
Fermentable organic matter	593.1
Microbial protein yield based on available energy	89.0
Microbial protein yield based on available nitrogen	191.6
Total protein supply to intestine	124.9
Absorbed microbial protein	56.7
Endogenous protein loss	11.8
Intestinal digestibility of rumem bypass protein (assumed)	85%
Absorbed rumen bypass protein in the intestine	49.4
Total truly and absorbed protein in the intestine	94.4
Degraded protein balance	102.7

Basic data used for nutrient modeling from Goelema (1999).

CONCLUSIONS

The molecular structure features of whole seed, hull and seed without hull could be generated using molecular spectroscopy. Functional groups and Functional group ratios differ among the different inherent structures (such as alph-helix to beta-sheet ratio). The structural difference at various regions could be detected using cluster analysis and principal component analysis. Peas have unique chemical composition which results in unique chemical ratio profile. Protein and carbohydrate subfractions are unqiue compared with cereal grain. Pea protein and starch have very high effective degradability which results in inefficient use in ruminants. Peas have high ratio of available N to available carbohydrates which is larger than the optimum ratio. The soluble or rapidly degradable protein is too high, resulting in unnecessary potential nutrient losses from the rumen. The heat processing could manipulate pea degradation kinetics to reduce rumen degradation and increase rumen bypass fractions. But the sensitivity of biopolymers to heating differs in pea protein and starch. Different heat processing methods also produce different heating effect. Modeling nutrient supply from peas shows that peas have 94 g/kg DM of total truly and absorbed protein in the intestine and 103 g/kg DM of degraded protein balance. More research needs to be done in detecting the sensitity of biopolymers or chemcial functional groups to the processing (such as heating, toasting). More research needs to be done in detect inherent structure changes in relation to nutrient availability.

ACKNOWLEDGMENT

This research has been supported by grants from the Ministry of Agriculture Strategic Research Chair Program. The author thanks Z. Niu and B. Liu for assistance in spectral data collection and analyses in my research team.

REFERENCES

Aguilera, J.F., Bustos, M., Molina, E., 1992. *The degradation of legume seed meals in the rumen: effect of heat treatment.* Anim. Feed Sci. Technol. 36, 101-112.

Cerning-Beroard, J., Filiatre, A., 1977. *Characterization and distribution of soluble and insoluble carbohydrates in legume seeds: horse beans, peas, lupines.* In: Protein Quality from Leguminous Crops. Commission of the European Communities, Belgium, pp. 65-79.

Czerkawski, J.W., 1986. *An Introduction to Rumen Studies.* Pergamon Press, Oxford, England, pp. 236.

CVB, Centraal Veevoeder Bureau, 1996. *Voorlopig Protocol Voor* In sacco Pensincubatie Voor Het Meten Van Eiwitbestendigheid, 14 Nov.

Doiron, K.J., Yu, P., McKinnon, J.J, Christensen, D.A. 2009a. *Heat-induced protein structures and protein subfractions in relation to protein degradation kinetics and intestinal availability in dairy cattle.* J. Dairy Sci. 92, 3319-3330.

Doiron, K.J., Yu, P., Christensen, C.R., Christensen, D.A., McKinnon, J.J. 2009b. *Detecting molecular changes in Vimy flaxseed protein structure using synchrotron FTIRM and DRIFT spectroscopic techniques: Structural and biochemical characterization.* Spectroscopy. 23, 307–322

Ensminger, M.E., Olentine, C.G., Jr. 1978. *Feed and Nutrition - Complete.* 1st Edn. Clovis, California, pp. 361.

Goelema, J.O., 1999. *Processing of legume seeds: effect on digestive behavious in dairy cows.* Ph.D Thesis. Wageningen Agricultural University, The Netherlands.

Himmelsbach, D.S., Khalili, S., Akin, D.E. 1998. *FT-IR microspectroscopic imaging of flax (linum usitatissimum L.) stems.* Cell. Mol. Bio. 44, 99-108.

Kemp, W. 1991. *Organic Spectroscopy.* 3rd ed. W.H. Freeman and Company, New York, USA.

Liu, N., P. Yu. 2010. *Using DRIFT Molecular Spectroscopy with Uni- and Multi-variate Molecular Spectral Techniques to Detect Plant Protein Molecular Structure Difference among Different Genotypes of Barley.* J. Agric. Food Chem. 58, 6264–6269

Marinkovic, N.S., Chance, M.R. 2006. *Synchotron Infared Microscopy, Encyclopedia of Molecular Cell Biology and Molecular Medicine,* 2nd Ed., Vol. 13, R. Meyers, Editor, Wiley Inc., pp. 671-708.

NRC. 1996. *Nutrient Requirement of Beef Cattle. 7th ed., National Research Council,* National Academic Press, Washington, DC, USA.

NRC. 2001. *Nutrient Requirement of Dairy Cattle.* 7th rev. ed., National Research Council, National Academic Press, Washington, DC, USA.

Pea: *http://en.wikipedia.org/wiki/Pea#Nutritional_value*; Retrieved 2011-08-23.

Petterson, D.S., Mackintosh, J.B., 1994. *The chemical composition and nutritive value of Australian grain legumes.* Grain Research and Development Corporation, Canberra, pp. 13-16.

Pownall TL, Udenigwe CC, Aluko RE (2010). *"Amino acid composition and antioxidant properties of pea seed (Pisum sativum L.) enzymatic protein hydrolysate fractions".* J. Agric. Food Chem 58, 4712–4718.

Samadi, P. Yu. 2011. *Determine heat-induced changes of protein molecular structure, protein subfraction and nutrient profiles in soybean seeds affected dry and moisture heat processing.* J. Dairy Sci. In press.

Sniffen, C.J., J.D. O'Connor, P.J. Van Soest, D.G. Fox and J. Russell. 1992. *A net carbohydrate and protein system for evaluating cattle diets: II Carbohydrate and protein system availability.* J. Anim. Sci. 70, 3562-3577.

Tamminga, S., van Vuuren, A.M., van der Koelen, C.J., Ketelaar, R.S., van der Togt, P.L., 1990. *Ruminal behavior of structural carbohydrates, non-structural carbohydrates and crude protein from concentrate ingredients in dairy cows.* Neth. J. Agric. Sci. 38, 513-526.

Tamminga, S., Van Straalen, W.M., Subnel, A.P.J., Meijer, R.G.M., Steg, A., Wever, C.J.G., Block, M.C., 1994. *The Dutch protein evaluation system: the DVE/OEB-system.* Livest. Prod. Sci. 40, 139-155.

Van Straalen, W.M., Tamminga, S., 1990. *Protein degradation in ruminant diets.* In: Wiseman, J., Cole, D.J.A., (Eds), Feed Evaluation. Butterworth, London, pp. 55-72

Walhain, P., Foucart, M., Thewis, A., 1992. *Influence of extrusion on ruminal and intestinal disappearance in sacco of pea proteins and starch.* Anim. Feed Sci. Technol. 38, 43-55.

Wetzel, D.L. 2001. *When molecular Causes of wheat quality are known, molecular methods will supercede traditional methods.* Proc. Int'l Wheat Quality Conf. II, Manhattan, Kansas, USA, May, pp 1-20.

Wetzel, D.L., Eilert, A.J., Pietrzak, L.N., Miller, S.S., Sweat, J.A. 1998. *Ultraspatially resolved synchrotron infrared microspectroscopy of plant tissue in situ.* Cell. Mol. Bio. 44, 145-167.

Yu, P., J.O. Goelema, B.J. Leury, S. Tamminga and A.R. Egan. 2002. *An analysis of the nutritive value of heat processed legume seeds for animal production using the DVE/OEB model: A review.* Anim. Feed Sci. Technol. 99 (1-4): 141-176.

Yu, P. 2004. *Application of advanced synchrotron-based Fourier transform infrared microspectroscopy (SR-FTIR) to animal nutrition and feed science: a novel approach.* Br. J. Nutr. 92, 869-885

Yu, P., McKinnon, J.J., Christensen, C.R., Christensen, D.A. 2004. *Imaging molecular chemistry of Pioneer corn.* J. Agric. Food Chem. 52, 7345-7352.

Yu, P. 2005. *Application of cluster analysis (CLA) in feed chemical imaging to accurately reveal structural-chemical features of feeds within cellular dimension.* J. Agric. Food Chem. 53, 2872-2880.

Yu, P. 2010. *Book Chapter 27. Detect Structural Features of Asymmetric and Symmetric CH2 and CH3 Functional Groups and Their Ratio of Biopolymers within Intact Tissue in Complex Plant System Using Synchrotron FTIRM and DRIFT Molecular Spectroscopy.* In: Biopolymers; ISBN 978-953-307-109-1; Editor: Professor Dr. Magdy M. Elnashar; pp 535-546.

Yu, P. 2010. *Plant-based food and feed protein structure changes induced by gene-transformation, heating and bio-ethanol processing: A novel synchrotron-based molecular structure and nutrition research program.* Molecular Nutrition and Food Research (MNF). 54, 1535–1545.

Yu, P. 2011. *Microprobing molecular spatial distribution and structural architecture of sorghum seed tissue (Sorghum Bicolor L.) with SR-IMS technique.* J. Synchrotron Rad. 18, 790-801

INDEX

D